Journalism and Clima

Journalism and Climate Crisis: Public Engagement, Media Alternatives recognizes that climate change is more than an environmental crisis. It is also a question of political and communicative capacity.

This book enquires into which approaches to journalism, as a particularly important form of public communication, can best enable humanity to productively address climate crisis. The book combines selective overviews of previous research, normative enquiry (what *should* journalism be doing?) and original empirical case studies of environmental communication and media coverage in Australia and Canada. Bringing together perspectives from the fields of environmental communication and journalism studies, the authors argue for forms of journalism that can encourage public engagement and mobilization to challenge the powerful interests vested in a high-carbon economy – 'facilitative' and 'radical' roles particularly well-suited to alternative media and alternative journalism. Ultimately, the book argues for a fundamental rethinking of relationships between journalism, publics, democracy and climate crisis.

This book will interest researchers, students and activists in environmental politics, social movements and the media.

Robert A. Hackett is Professor of Communication at Simon Fraser University, Vancouver and a co-founder of NewsWatch Canada, Media Democracy Days and OpenMedia.ca. His research interests include media democratization; critical news analysis; and social movements, peace and media. Recent collaborative books include *Expanding Peace Journalism* and *Remaking Media*.

Susan Forde is Director of the Griffith Centre for Social and Cultural Research and Associate Professor of Journalism at Griffith University, Australia. Books include *Challenging the News* and *Developing Dialogues* (with Meadows and Foxwell). Until 2002 she was a journalist in the alternative, independent and Indigenous media, including editing the Indigenous newspaper *Land Rights Queensland.*

Shane Gunster is Associate Professor in the School of Communication at Simon Fraser University and a research associate with the Canadian Centre for Policy Alternatives. His teaching and research interests focus upon news media coverage, advocacy and engagement around the politics of energy and climate change.

Kerrie Foxwell-Norton is Senior Lecturer in Journalism and Media Studies at Griffith University, Australia. Her research investigates the role of local media and communities in addressing global environmental issues, and her key interest is in the intersections between media, policy and the environment. She is the co-author of *Developing Dialogues* (with Forde and Meadows).

Communication and Society

Series Editor: James Curran

This series encompasses the broad field of media and cultural studies. Its main concerns are the media and the public sphere: on whether the media empower or fail to empower popular forces in society; media organizations and public policy; the political and social consequences of media campaigns; and the role of media entertainment, ranging from potboilers and the human-interest story to rock music and TV sport.

For a complete list of titles in this series, please see: https://www.routledge.com/series/SE0130

News and Politics
The Rise of Live and Interpretive Journalism
Stephen Cushion

Gender and Media
Representing, Producing, Consuming
Tonny Krijnen and Sofie Van Bauwel

Misunderstanding the Internet
Second Edition
James Curran, Natalie Fenton and Des Freedman

Africa's Media Image in the 21st Century
From the 'Heart of Darkness' to 'Africa Rising'
Edited by Mel Bunce, Suzanne Franks and Chris Paterson

Comparing Political Journalism
Edited by Claes de Vreese, Frank Esser and David Nicolas Hopmann

Media Ownership and Agenda Control
The Hidden Limits of the Information Age
Justin Schlosberg

Journalism and Climate Crisis

Public Engagement, Media Alternatives

Robert A. Hackett, Susan Forde,
Shane Gunster and
Kerrie Foxwell-Norton

Routledge
Taylor & Francis Group
LONDON AND NEW YORK

First published 2017
by Routledge
2 Park Square, Milton Park, Abingdon, Oxon OX14 4RN

and by Routledge
711 Third Avenue, New York, NY 10017

Routledge is an imprint of the Taylor & Francis Group, an informa business

British Library Cataloguing-in-Publication Data
A catalogue record for this book is available from the British Library

Library of Congress Cataloging-in-Publication Data
A catalog record for this book has been requested

ISBN: 978-1-138-95038-2 (hbk)
ISBN: 978-1-138-95039-9 (pbk)
ISBN: 978-1-315-66873-4 (ebk)

Typeset in New Times Roman
by Apex CoVantage, LLC

Contents

Figures

Acknowledgements

A book of this scope, with four co-authors in two continents, would not have been possible without many contributors. We are indebted to Michael Meadows, now retired from Griffith University, for bringing us together in the first place.

We gratefully acknowledge the financial support of Griffith University through the Griffith-SFU Travel Collaborative Grant program which enabled Forde and Foxwell-Norton to visit Vancouver in 2014. Visits by Hackett and Gunster to Griffith University to work on the manuscript in 2016 were fully financed by the Short Term Visiting Research Fellowship Scheme through the Arts, Education and Law Group. Additional funding for data collection and the preparation of the related chapters was also provided by Griffith University's Centre for Social and Cultural Research and the former School of Humanities.

Much of the empirical research in British Columbia was supported through the Climate Justice Project of the Canadian Centre for Policy Alternatives (CCPA) and the University of British Columbia, funded by the Social Sciences and Humanities Research Council of Canada (SSHRC) – particularly, a project/module entitled 'The climate of discussion: News media and the politics of climate change in British Columbia,' led by Kathleen Cross with Shannon Daub, Gunster and Hackett as co-investigators, and with research assistance from Marcelina Piotrowski, Sibo Chen, Helena Krobath and Madison Trusolino. Particular projects/modules also received funding from the SSHRC, and from the work/study program, the Vice-President Research, the Dean of Graduate Studies, and the Dean of Communication, Art and Technology, at Simon Fraser University.

Hackett acknowledges research assistance from undergraduate students at SFU. Over several years, Ashley Adams, Sahib Bhatia, Joey Chopra, Mariam Faqeri, Kathryne Gravestock, Rohini Grover, Kavya Joseph, Nicole Keith, Kevin Keyhoe, Wendee Lang, David Lazenby, Vojtech Sedlak, Josh Tabish, Maegan Thomas, Sara Wylie and Rebecca Visser helped to keep pace with the burgeoning research literature in alternative media, media democratization and environmental communication. Foxwell-Norton acknowledges research assistance from Terri Lethlean for the data gathered in Chapter Six. The Australian authors also acknowledge (and utilized) the comprehensive literature research work from the Canadian-based research assistants working with Hackett.

For permission to use portions of his previously published articles, Hackett thanks his co-authors Sarah Wylie and Pinar Gurleyen and journal editor Richard Keeble ('Enabling environments: reflections on journalism and climate justice', *Ethical Space: The International Journal of Communication Ethics* 10(2/3), 2013); Irena Regener, publisher of *Conflict and Communication Online* and *Peace Journalism: The State of the Art*, edited by Dov Shinar and Wilhelm Kempf ('Is peace journalism possible?', 2006); and Stuart Trew, editor of the CCPA *Monitor* ('Media reform and climate action', July/August 2016).

Directly and indirectly, we have benefitted from the advice of our colleagues – in Vancouver, including Frank Cunningham, Andrew Feenberg, Katherine Reilly, Yuezhi Zhao – and respondents at conferences of the International Environmental Communication Association and International Association for Media and Communication Research, where Gunster and Hackett presented previous versions of their chapters. Gunster thanks participants in his environmental communications graduate seminar over the past five years who provided a fertile and stimulating intellectual environment for developing many of the arguments in Chapter Two, and Chris Russill and Roy Bendor for helpful comments on its earlier draft. In Australia, Foxwell-Norton and Forde thank research colleague Michael Meadows for his advice and input, and the wide range of scholars, coastal community groups and representatives of ENGOs that have contributed to our annual Communication4Conservation symposium, which has provided motivation for much of the material in Chapters Three, Six and Seven.

We are indebted to Angelika Hackett for extensive editorial assistance on the entire manuscript and references, and to our patient and helpful editors at Routledge, Natalie Foster and Sheni Kruger. And of course, the project could not have been completed without the support of our respective partners and families. We hope that our contribution on a topic so fundamentally affecting their future has justified their forbearance.

Introduction

Journalism(s) for climate crisis

Robert A. Hackett, Susan Forde,
Shane Gunster and Kerrie Foxwell-Norton

> *To be truly radical is to make hope possible, rather than despair convincing.*
> (Raymond Williams, *Resources of Hope*)

Ordinary journalism in extraordinary times

As this book nears completion, a massive wildfire ravages the Canadian city of Fort McMurray, epicentre of one of the largest and most controversial energy mega-projects in the world – the Alberta tar (or oil) sands, containing billions of barrels of bitumen whose full exploitation would unleash both enormous revenues and further global warming. But after weeks of exceptional heat and aridity, the wildfires forced more than 80,000 residents to flee their homes, one of the largest evacuations in Canadian history. While a few politicians, environmentalists, scientists and journalists cautiously raised the connection between the wildfire and climate change, many – including Canadian Prime Minister Justin Trudeau – not only refused to draw these linkages but also questioned those who suggested them (Dann 2016; De Souza 2016). Even fewer dared to point to the tar sands' complicity in the very process that is uprooting so many lives and jeopardizing its own infrastructure.

On the other side of the planet, devastating wildfires have raged in Australia with increasing frequency and intensity (Hughes and Steffen 2013). In 2009, 173 lives were lost in the 'Black Saturday' bushfires that destroyed an entire community. Bushfires are common in Australia and have occurred throughout history – but evidence is firm that hotter temperatures and longer fire seasons are increasing the incidence and intensity of fire, particularly in those areas already susceptible (Hughes and Steffen 2013: i–ii). Yet as late as 2013, then-Prime Minister Tony Abbott criticized a United Nations official for 'talking through her hat' for linking the fires with climate change (Ireland 2013). And wildfires are not the only 'natural' disaster associated with climate change. The Great Barrier Reef off the Queensland coast in the north-east of Australia is undergoing one of the most significant bleaching events in recent memory, a consequence of warming oceans and directly (if not exclusively) connected with climate change, according to climate and coral reef scientists (Slezak 2016b). Frustrated by the ongoing failure of Queensland's largest daily – Rupert Murdoch's *Courier Mail* – to provide accurate coverage, dozens of those scientists felt compelled to place an advertisement in the paper in 2016 to highlight the magnitude of the problem and its connection to the mining, export and burning of coal (Slezak 2016a).

It seems that in both countries, the boundaries of public discourse about climate change are being constrained by powerful sections of the media and political establishments.

The reality of anthropogenic climate change can no longer be ignored. But neither can the responsibility of journalism to inform, motivate and empower citizens to engage with the problem. Climate change is a crisis in a dual sense (Cottle 2009a). Global warming's disruption of nature's metabolism has a biophysical reality that threatens the fundamental preconditions of human habitats. But it acquires presence as a crisis only when it is discursively represented as such, entailing a demand for urgent attention and an insistence that the status quo cannot continue without undermining its own foundations. As Ulrich Beck's concept of the risk society suggests, climate crisis is not only about ecological decline but also our political and communicative capacity to respond, including the 'key social and political position' of mass media in the social definition of unfathomable global risks (Beck 1992: 22–3).

The exigencies of that response place demands not only on overtly political institutions, but also on journalism as one of modernity's most important forms of storytelling. While its sites and sources are multiplying and its conceptual boundaries are blurring in the digital age, journalism as an institutionally based cultural and political practice continues to have an active role in defining and addressing crises. Its recognized functions include setting agendas for publics and policymakers, helping to define social problems, certifying issues and events as publicly relevant and 'fixing' their meanings, constructing hierarchies of access to the public realm, participating in the construction or mediation of subjectivity, legitimating or challenging established institutions and ways of thinking, helping to create or undermine communities both virtual and geographical, and influencing the trajectory of social movements (see e.g. Benson *et al.* 2012; Downing *et al.* 2001; Franklin 2012; Hallin and Mancini 2004; McChesney 2003; Zelizer 2011, among many others).

For such reasons, journalism plays a key role, alongside other institutions and movements, in shaping more effective responses to climate crisis. And yet, the conventional news media – corporate- or state-owned broadcasters and press and their extensions on the Internet – are clearly falling short of the needs of the times. Indeed, such media are themselves in crisis; once 'hegemonic', in the sense of both dominating their market and generally reproducing ideologically dominant definitions of reality, their erstwhile dominance is under severe challenge in the Internet era. But conventional media (now often referred to as 'legacy media') are still a primary means by which people orient themselves towards the world beyond their direct experience and, in particular, make sense of their own capacity for individual and collective agency. Importantly then, media scholars and critics have identified key 'deficits' in news coverage of climate change.

In part, the problem is simply the sparseness of coverage, and of resources devoted to climate change and climate politics. With some exceptions, conventional journalism falls vastly short of confronting the causes, scale, consequences, urgency and complexity of the challenge. In the heavily populated parts of the world most affected by the impacts of climate change – countries in the global South, at risk from extreme storms, drought or rising sea levels – news institutions often lack resources and journalists 'struggle to report effectively on climate change due to a lack of training, unsupportive editors and weak outreach from domestic policymakers' (Climate Change Media Partnership 2011). Too often, journalism overlooks marginalized people, vernacular languages and traditional knowledge, and presents climate change as international news without local relevance. There are signs of recent improvement, such as the increased employment of science journalists, in the global South (Boykoff and Yulsman 2013: 363). But news about global processes and policy forums continues to be dominated by hegemonic media organizations that are oriented towards the affluent audiences of the global North (a colloquialism that includes economically 'advanced' countries in the southern hemisphere, like Australia) and 'the viewpoints of the more powerful nations and international nongovernmental organizations' (Climate Change Media Partnership 2011).

Inadequate reporting of climate crisis is hardly confined to the global South. In the world's largest economy, American publics have remained confused about the extent of scientific consensus concerning the human causes of climate change. That confusion is largely due to systematic disinformation by the vested interests that would stand to lose from a lower-carbon economy, and from the hegemonic media's practice until recently of 'balancing' climate science with climate deniers (Boykoff and Boykoff 2004; Oreskes and Conway 2010, Chapter 6). Simon Cottle, a leading researcher on media and global crisis, sees a 'fundamental disconnect between the media's representations of climate change and the politics and policies needed to effect meaningful change' (2009b: x).

However – and this is a theme of our book – *the key problem is not a lack of coverage or information.* The key questions are *how* journalism presents the issue and what kinds of responses it generates in audiences/publics. Hegemonic media's focus on disaster, threat and elite political squabbling generates a 'hope gap' that leaves audiences, even those most alarmed by climate change, with a sense of disengagement and powerlessness, rather than efficacy (Upton 2015). A dozen climate change communicators working in alternative media or environmental advocacy groups were interviewed for the Climate Justice project in Vancouver, Canada (see Chapter Five). Those interviews generated many complaints, beyond a lack of coverage, that are broadly supported by academic research. In their view, coverage is episodic and compartmentalized. It too infrequently connects the dots between, for example, the manifestations of climate change and its causes and consequences, or the rapid exploitation of fossil fuels and global warming. News routinely relies upon a narrow range of official sources (in our respondents' view), and overemphasizes disasters, rather than positive change agents and creative solutions. When solutions are addressed, there is too much focus on technology and individual 'green consumerism' rather than collective approaches and policy options. The overall editorial environment favours economic growth, consumerism and private-sector business. There is little attention to who bears responsibility for climate change, and little critical analysis of capitalism or even the fossil fuel industry. Climate change is sometimes framed as inevitable, generating a sense of defeatism. Some respondents regard media focus on environmental conflicts as unproductive (although we argue in Chapter Five that the way conflict is framed is critical, and some frames can enhance popular mobilization).

Many of the foregoing complaints from our respondents are supported by systematic studies of news content, and are rooted in the political economy of news media and professional ideologies of conventional journalism (e.g. Boykoff and Yulsman 2013; McChesney and Nichols 2010). Indeed, our respondents are reflecting the persistent concerns of alternative media practitioners and alternative journalists (see Atton and Hamilton 2008: 70ff; Forde 2011: 56ff; Harcup 2013: 128ff), albeit this time in the context of reporting environmental issues and climate change. Increasingly, we hear of a 'crisis of journalism', especially in the U.S. and other advanced democracies such as the U.K., Australia and parts of Europe (Franklin 2012: 597–8). Worried scholars and journalists point to an apparently collapsing business model, as audiences for conventional media decline and fragment, and advertising revenue stampedes to digital media. The 'millennial' generation of digital natives largely ignores conventional news, and few people of any age are willing to pay much for it. And the conglomerates that increasingly dominate media ownership are maximizing short-term profits, stripping assets and disinvesting in news (Freedman 2010; McChesney and Nichols 2010). Working journalists are faced with tighter deadlines, heavier workloads, multi-platform demands, a 24/7 news hole to fill and a broader palette of topics to report. The result is predictable: fewer beat reporters with specialized expertise, less investigative or accountability journalism, more pressure to act like stenographers, reporting competing claims rather than assessing their respective validity (McChesney 2003).

Beyond the crisis of journalism's business model, however, Climate Crisis Journalism faces additional barriers of institutional structure, class power and ideology that go well beyond conventional economists' notions of market failure. As Naomi Klein (2014) argues, taking global warming seriously requires a positive role for government, a strengthened public sector and collective action – which is precisely why conservative political forces, especially in North America and Australia, prefer *not* to take it seriously. A similar ideological animus might apply to commercial media. After all, ecologically destructive capital logic fuels such media's need to attract profitable upscale audiences, media's imbrication with the growth of the urban middle class in global capitalism's 'emerging economies' such as China and India, and media's project of colonizing popular imagination with consumerist lifestyles. Almiron (2010) sees a qualitative leap in the recent integration of news media with contemporary capitalism. As finance capital comes to dominate the industrial sphere, corporate media prioritize financial information and services at the expense of journalism, and become speculative actors themselves, desperate to increase profits and revenues (Boyd-Barrett 2011). The growing concentration and globalization of news media ownership, and the expansion of a global public relations industry with sophisticated media strategies, are further structural barriers that yield an emphasis on human interest, celebrity and entertainment-oriented reporting, at the expense of complex multifaceted issues (Anderson 2009: 178–9). Advertising helps to create 'a set of cultural conditions that makes us less inclined to deal with climate change', so that 'a media and telecommunications industry fuelled by advertising and profit maximization is, at the moment, part of the problem rather than part of the solution' (Boyce and Lewis 2009: 8–9).

The evolving mediascape of online commercial journalism does not promise much better, despite the technical potential of the Internet for explanatory and solutions-oriented journalism. Editorial decision-makers can now instantly determine what type of stories attracts the most 'click-throughs' – the most 'unique visitors'. Future content is influenced by running stories that will maximize 'clicks', typically celebrity news and sensational statements, rather than more substantial news relevant to democracy and political efficacy.

Journalism that is suborned by such imperatives will likely be muted on an issue that implicitly evokes the need for collective action that bursts the constraints of market relations, consumerism and property rights.

It is important not to paint too bleak and monolithic a picture. Excellent climate journalism thrives in some non-profit segments of the global media; the U.K.-based *Guardian* is an outstanding example and is featured in Chapter Seven. It can be found even in corners of conventional commercial media. In southern California, stricken in recent years by extreme drought, Palm Springs's daily paper, *The Desert Sun*, has produced a series of striking feature articles on regional aspects of climate change. Written by its environment and energy beat reporters Ian James and Sammy Roth, and accompanied by dramatic photographs, the series addressed not only various biophysical impacts, but also climate politics, such as a respectful portrayal of the 'Keep it in the Ground' anti-fossil-fuel movement, a piece rebutting the argument that environmental protection kills jobs, and a feature on where the presidential candidates stand. In 2014, *The Desert Sun* made an intervention relatively unusual for a corporate news organization, co-sponsoring a free public forum on the region's water crisis, reinforced by extended background articles.

Another example of productive coverage is offered by the *Burnaby Now*, a semi-weekly 'community newspaper' in the hometown of two of this book's co-authors. Since 2014, the paper, particularly reporter Jennifer Moreau, extensively covered the controversy surrounding the proposed expansion of a pipeline to facilitate export of Alberta's bitumen reserves. Owned by Texas-based Kinder Morgan, the pipeline would slice through the coastal British Columbia city and sharply increase the oil tanker traffic plying the region's unceded aboriginal territories and environmentally sensitive harbours. In its pages and website, the paper offered different perspectives on the proposal, including respectful and human-interest treatment of the widespread community opposition to it. Similar examples can be drawn from the Australian context and throughout the world, where the 'beat' is a local community or geography and news coverage of threats or risks to local places is regular reporting (see Foxwell-Norton 2015).

Such journalism is arguably made possible by the initiative of particular journalists and news organizations, in the context of enabling market conditions and/ or citizen action through 'social media' or more traditional avenues – such as the opposition of a majority of Burnaby's residents and elected representatives to the pipeline, or Palm Springs's economic dependence on tourism and renewable energy (rather than fossil fuel extraction). In the conventional, large-scale media, however, it seems all too rare, given the structural pressures just sketched. The contrast between the urgency of public and political action on climate crisis and the deficits of journalism about it have catalyzed this book. Although good practice is identified and presented in later chapters, we begin with a broader normative enquiry. In the context of climate crisis, what could be the relationship between journalism, publics and politics? What ethos, priorities or ethical guidelines should inform it? In addressing such questions, we are mindful of the landmark work by Clifford Christians and his colleagues on normative theories of the media. In developing concepts of the 'ideal' media, they argued, normative theory requires a standpoint from which to evaluate the status quo, not merely describe it, and to explain and inspire a transition from 'what is' to 'what ought to be' (Christians *et al.* 2009: 92–3). On the other hand, if it is to be useful as a guide to ethical, structural and practical change, then critical normative theory cannot be too far divorced from current realities.

Thus, the reforms or standpoints for journalism proposed in this book are informed not only by previous normative theorizing on democracy and the public sphere, for example, but also by empirical research on journalism content and practices, alternative media, environmental communication and social movement strategies. Some of the essays report the co-authors' own original research from Australia and Canada – two countries with similar political systems, histories of colonization vis-à-vis Indigenous peoples, extractivist and export-oriented economies, and formally independent (of government) but heavily concentrated media ownership. Our empirical cases include essays on the views and strategies of environmental communicators and alternative journalists in Vancouver regarding climate crisis communication, the performance of a major online news medium in Australia with regard to the 'Keep it in the Ground' campaign, and the continuities and contrasts in conventional and alternative media's climate change news in Australia. Interviews, focus groups, textual readings and content analyses are the main methods employed. That empirical work, along with overviews of previous scholarship, informs our recommendations for media change. The theoretical perspectives are eclectic, reflecting the authors' diverse backgrounds in journalism studies, environmental communication, political science, community and alternative media, and communication studies, as well as direct involvement in alternative journalism, Indigenous media, media democratization and community struggles over energy politics and coastal policy.

Themes of the book

Although Hackett acted as coordinating author, and while there was collegial exchange of views and sometimes text between the four of us, the essays in this volume reflect each author's individual views. The book as a whole coalesces around several cross-cutting themes and commitments.

First, in light of our own empirical findings and environmental communication research more broadly, we argue that the key deficit of conventional media (at least in countries such as Australia, Canada, the U.S. and the U.K.) is *not* a shortage of information. If there is a lack of political and policy action to mitigate climate change, notwithstanding the scientific consensus on the broad question, it is not because people in general do not know enough about it. We cannot assume that simply providing people with accurate information about its causes, consequences and solutions would automatically generate greater concern and support for policy solutions and behavioural change. Rather, the key deficit is one of agency, hope and efficacy. Millions of people are concerned about climate change, but they do not know what to do about it. Indeed, they often feel that their own contribution to it may not be worth the effort – if no one else is doing it, why should I? We maintain that the necessary action of dominant political and economic elites is compromised by their vested interests and the privileged conditions of their existence. Hence, public engagement and widespread popular political mobilization are critical to achieve the fundamental changes – in the economy, energy policy, urban planning, trade policy, 'lifestyles', cultural narratives – needed to check climate change's worst impacts. Yet many people lack faith in political institutions of all kinds, in other people's motives, and in the possibility of collective action.

Thus, we argue that in covering urgent political issues such as climate change, journalism needs to re-think its mission. It's less about the *informed* citizen, and more about *engaged* publics. The task is not so much making climate change real for people, but rather enabling a sense of efficacy, a belief that individual and collective action is possible and potentially effective, and conveying a sense of crisis and urgency without overwhelming people with the scale and pace of the challenge.

There is much in the work of scholarship on alternative media and, specifically, alternative journalism which adds ballast to this aim. Chris Atton's (2002) work on alternative media draws on historical studies of the alternative press to disclose 'methods of production and distribution, allied to an activist philosophy of creating "information for action" timeously and rapidly' (p. 12). He identifies alternative media as 'oppositional in intent, having social change at their heart' (p. 19). Tony Harcup notes the involvement of ordinary people in producing alternative journalism as a form of 'active citizenship' which by definition implies active engagement in society, and involvement 'in some form of collective or political endeavour' (2013: 140). He concludes that 'it must be hoped that alternative forms of journalism will not continue to be seen as of marginal importance whenever the relationship between journalism and democracy is discussed and analysed' (p. 141).

Harcup's connection of active citizenship and alternative journalism is both important, and consistent with much scholarship that identifies the empowerment that *production of* and *participation in* such media delivers (e.g. Atton 2002; Atton and Hamilton 2008; Hackett and Carroll 2006; Rodriguez 2001). This is a more common theme than the ways in which alternative media news content might activate audiences *outside* the media outlet, although John Downing (2001) notes that political and social change is the *primary purpose* of radical journalism. Collins and Rose (2004) recount their experience establishing the alternative *City Voice* in Wellington, New Zealand, which 'tried with public journalism to empower people to understand issues and to actually do something about them' (p. 34). Susan Forde's first survey of Australian alternative press journalists in 1996 found 'motivating the public' as one of their primary journalistic aims (1997: 118). The literature about Public Journalism or Civic Journalism, discussed in Chapter Four, certainly identifies the need for the media to create a forum in which people can discuss, engage and interact and where they are empowered to act, although it has since been recognized that while the normative ideal of public journalism could potentially reform audiences as 'active citizens', the reality of embedding public journalism in mainstream, corporate media is a somewhat failed 'patrician' experiment (Atton and Hamilton 2008: 64; Tanni Haas's extensive work (2004) on public journalism is also reluctantly sceptical about its impact; Haas and Steiner 2006).

Closer to the ideal is Jankowski's discussion of community media, which he found focused on providing news and information relevant to the needs of the identified community 'to engage these members in public discussion, and to contribute to their social and political empowerment' (2003: 4). Forde (2011) reports on alternative journalists from the U.K., the U.S. and Australia who see 'activating the public' as one of their key roles in society. Such journalists were driven by the need to provide context to news already covered in the mainstream, and to provide information to their audiences which will overtly encourage them to take part in democracy, 'to participate, to *do something*' (p. 165; emphasis in original).

This book, then, also builds on previous assessments and theorizing of alternative journalism around notions of activating and engaging publics. We take this further to sharpen the motivation and intention of journalists reporting climate crisis, which is to use the *probable outcome* of their journalism as a starting point for the practice. While it is not possible to predict the outcome or impact that the reporting might have, it is possible to conceptualize stories, plan research, structure the news-gathering process, and execute story-writing with the *probable outcome* – indeed, the *desired outcome* – as a guiding principle. This implies a journalism which is designed to foster not only awareness and understanding, but also a sense of efficacy, agency, and power in the audience.

This raison d'être of alternative media, and particularly alternative journalists in all their guises, provides a substantial touchstone for this book. Its second distinct theme, then, is the actual and potential roles of 'alternative media' in relation to climate crisis. Since Downing's landmark study of radical media (2001), scholarship in this field has burgeoned, but relatively little has focused on environmental concerns. We note Brett Hutchins and Libby Lester's (2015) work on the 'switching points' that explain mediatized environmental conflict, including environmental activists and their media. In Malaysia, Sandra Smeltzer (2008) distinguished between mainstream state-controlled media, and online alternative media that fill a gap in informing citizens about environmental issues. In El Salvador, Hopke (2012: 377–8) has traced the way in which alternative media there supported the anti-mining social movement, acting as 'counter-narrative', elevating community interests of human rights and environmental justice; and there were earlier considerations of the role of progressive media in conveying environmental issues (see e.g. Atton's (2002) sections on the U.K.'s *Green Anarchist*).

Even fewer studies focus on alternative media's contributions to climate crisis communication as such; Linda Jean Kenix (2008) may have been the first, followed by Foxwell-Norton (2015) and Gunster (2011, 2012). Given the potential affinity of the recognized characteristics of alternative media, including oppositional content and links with communities and social change movements, that gap is surprising. We attempt to address that gap, but recognize immediately a problem of nomenclature. A variety of cognate terms have been used by different authors, including alternative, alterative, autonomous, citizens', community, critical, independent, native, participatory, progressive, radical, social movement and tactical media. Arguably, different politico-theoretical traditions are attempting to appropriate the mantle of alterity and catalytic change (see e.g. Hadl and Dongwon 2008). Often the debate focuses around the question of whether alternative media are defined by their (oppositional) content, or their (participatory) production processes (see e.g. Fuchs and Sandoval 2015). Alternative media, as a rule, present information focused on outcomes, public action and engagement. Furthermore, such media sometimes critique dominant journalism practices and how those practices might actually contribute to the *repression* of goals and objectives that could activate the citizenry. For our purposes, media are 'alternative' when they stimulate productive popular engagement and a sense of efficacy and empowerment on the issue of climate crisis.

To be sure, the conceptual dichotomy of 'alternative' and 'mainstream' media risks overlooking hybridization, reciprocal influence and oppositional moments within dominant media (e.g. Kenix 2011). Clemencia Rodriguez (2001) is particularly concerned about that binary juxtaposition, which focuses on what alternative *is not* rather than what it *is*. We are primarily interested in alternative *journalisms*, which are defined by their logics and outcomes more than their particular online or institutional settings, which can vary. Still, 'alternative' media's ability to nurture frames and paradigms suited to addressing global crisis is enhanced by many of their ideal-typical characteristics – participatory production, horizontal communication, openness to social movements, localism and engagement with communities, and independence from state and corporate control. One purpose of this book is to highlight our collective research on how alternative and community media can promote public engagement and critical perspectives specifically on environmental issues, and to encourage further exploration of this issue.

A third theme of the book is indicated by the plural in its subtitle: Public Engagement, Media Alternatives. Notwithstanding our commitment to building alternatives, we approach the task of exploring Climate Crisis Journalism with an openness to different options. No single type of journalism could meet all the demands of climate crisis – or for that matter, the demands of democratic communication (Curran 2002). In surveying the literature on media and democracy, Christians *et al.* (2009: 125) identify four key roles that are expected of journalism, emerging from 'conflicting requirements and value positions' in a complex political, cultural and media environment. We have used these as a touchstone in several of the chapters to emphasize the need for diverse forms of journalism – options and alternatives. We now outline them.

Four democratic roles for journalism

Christians and his co-authors (2009) took on the unenviable task in the first decade of the 2000s to update Siebert, Petersen and Schramm's (excessively) influential 1956 work, *Four Theories of the Press*. While authors such as Daniel Hallin and Paolo Mancini (2004) have made substantial contributions to describing and categorizing media systems in recent years, it is the Christians *et al.* work that takes on the more specific task of revising and updating normative theories of the press. They first describe four important 'traditions' in the history of public communication, but their identification of four roles of the media in democratic societies is the primary concern here.

By way of entry to the different 'genres' of journalism that we might identify, there is a growing arm of journalism studies which investigates those practitioners and institutions carrying out journalism within a different framework. Barbie Zelizer (2011: 9) notes that globally, there are 'multiple journalisms evident on the ground', and communications scholars must not ignore the diversity that does exist:

> certain privileged forms of journalism – the very notion of a free and independent press; the idea of a fourth estate or the public's right to know; and the embrace of neutrality, facticity, and objectivity – were never the practice in much of the world . . . today multiple modes of journalistic practice underscore how divergent and open-ended the field needs to remain, how sensitive to different and often contradictory cores, values, and contingencies, how relative and particularistic.
>
> (2011: 12–13)

Essentially, what is understood and accepted as journalism has many different forms around the world. Even within a democratic political system and its prevailing normative commitment to fourth-estate watchdog journalism, we can only fully understand and therefore evaluate 'journalism' if we acknowledge its many different faces. Consistent with this recognition, we summarize the four modern roles for the media in democratic societies – monitorial, facilitative, radical and collaborative – that Christians *et al.* outline.

The role of *monitoring* and reporting on current and recent events, and scanning the social horizon for future developments, could include critical scrutiny of abuses of power – a watchdog function – but does not extend to partisan advocacy (Christians *et al.* 2009: 125). This reportorial role is often coupled with notions of accuracy, fairness and *objectivity*, considered to be a hallmark of professional journalism in many advanced democracies for much of the past century.

Because monitoring is the 'most widely recognized' of journalism's democratic roles (2011: 12–13), and because climate crisis increases the stakes of longstanding debates about objectivity, we give it particular attention here. Objectivity is not a single construct, but a multifaceted discursive 'regime', a complex of ideas and practices that provides a general model for conceiving, defining, arranging and evaluating news texts, practices and institutions (Hackett 2008: 3345; Hackett and Zhao 1998). It has normative, practical, epistemological, institutional and discursive dimensions, and given its malleability and complexity, it does not have the same purchase in all countries, even in the global West. Western Europe's relatively vibrant and pluralistic press system has a strong tradition of partisanship. Moreover, 'objectivity' can have different meanings in practice, such as the negation of journalists' subjectivity, the fair representation of each side in a controversy, balanced scepticism towards all sides in a dispute, and the search for hard facts that can contextualize a dispute (Donsbach and Klett 1993).

Thus, contradictory practices and epistemological positions can claim justification as 'objective'. For example, objective monitoring could reasonably entail that journalists accept the scientific consensus on global warming's link with greenhouse gas emissions as a basis for selecting appropriate and credible sources, investigating its impacts and reporting on proposed solutions. By contrast, objectivity has sometimes been taken to imply a different attitude towards 'truth' – that it is to be discovered not through uncovering the facts of the case, but rather through splitting the difference (or taking no position at all) between two partial and presumptively biased accounts. In climate news, it led elite media in the U.S., Australia and elsewhere to grant equivalence to valid knowledge based on science, and denial based on opinion and vested interests (Boykoff and Boykoff 2004). The impact of such false balance in generating public confusion and delaying effective policy responses to climate crisis has been well documented.

Scholars of journalism have identified other biases of 'objective' reporting, including over-dependence on official sources; journalists' blindness to the frames that give meaning to stories; stereotypical constructions of conflict as two-sided zero-sum games, to the exclusion of multiple stakeholders and win-win solutions;

and a focus on today's events (often 'pseudo-events' staged for the purpose of being reported) rather than long-term conditions, contexts or processes such as global warming (Lynch and McGoldrick 2005: 209–12; Maras 2013). How do media users respond to such biases? There is a need for more research on this question, but experiments in other forms of journalism (discussed in Chapter Four) suggest that, notwithstanding the well-meaning intentions of journalists, some of the various practices of objectivity can reinforce tendencies towards feelings of political cynicism, disengagement, disempowerment and relativism – allowing citizens to believe what they choose (Bennett 2009: 211). Thus, 'objective' monitorial journalism may well favour some kinds of narratives, perspectives and outcomes over others, in ways that are not conducive to public engagement on climate crisis. That point should be kept in mind before dismissing other journalistic roles that are more explicitly linked to public objectives, such as facilitating or radicalizing civil society.

The *facilitative* role entails a conscious mission to improve the quality of public life, to promote active citizenship through broadened public debate and participation, and to enhance inclusiveness, pluralism, and collective purpose (Christians *et al.* 2009: 126). Civic democracy provides the logic and rationale for the media's facilitation of public life (p. 158). It harks back to James Carey's notion that journalism *only* makes sense in relation to the public, whereby 'the media promote dialogue among their readers and viewers through communication that engages them and in which they actively participate' (Christians *et al.* 2009: 126). The facilitative role sees reporters and editors supporting and strengthening participation in civil society 'outside the state and market . . . the media do not merely report on civil society's associations and activities but seek to enrich and improve them' (Christians *et al.* 2009: 126). Importantly, and further, citizens are taken seriously as actors who can clarify and resolve public problems in this iteration of the media's role, which is often linked to notions of the public sphere – 'a neutral space within society, free of both state or corporate control, in which the media should make available information affecting the public good, and facilitate a free, open and reasoned public dialogue that guides the public direction of society' (Curran 2000: 135). This characterization of the media in society sits closely with our position that journalism should be focused on its probable and, indeed, desired outcomes, by contrast with the monitorial role that may impotently deliver information or unintentionally generate negative outcomes, such as political cynicism. The facilitative role openly identifies a public engagement function for the media, and sees the public as important actors in addressing public problems and, presumably, in fostering social change as a result.

The *radical* role for the media is historically important but less widely recognized in conventional textbooks. It foregrounds social injustice, inequalities, abuses of power and the potential for political and social change, often in opposition to established authorities and power relations, and in sympathy with social movements representing marginalized or disenfranchised groups. Radical journalism aims to eliminate concentrations of social power to achieve true equity and fair participation in decisions affecting society. Beyond encouraging the public engagement and action that the facilitative role evokes, journalists fulfilling their radical role encourage 'changes in the core of existing social institutions . . . the long-range goal is a society of universal recognition of human rights for all' (Christians *et al.* 2009: 179). Radical journalists support change in the systems of communication so that the under-represented and the dispossessed receive all the information they need and can articulate 'an alternative set of goals that represent the needs and just moral claims of all' (Christians *et al.* 2009: 179). While Christians *et al.* assert that this role 'is not inconsistent with professionalism' (p. 126), it would likely make many adherents of 'professional' forms of journalism uncomfortable with the level of advocacy it could entail. We nevertheless argue in this book that space for this genuinely radical approach is required in reporting climate crisis, and that the resulting content should reach a broad and diverse audience.

Less relevant to this study but accurately detailed by Christians *et al.* is the *collaborative* role. The collaborative role specifies journalism's support for broader and dominant social purposes, from dealing with crises and emergencies, to promoting economic and social development – especially in postcolonial societies facing political instability, institutional fragility and resource scarcity (p. 127). In principle, journalism could collaborate with civil society, including movements for social change; but most Western theorists focus on relations with the state, as allegedly the only institution that 'can intervene in the affairs of journalism in ways that fundamentally alter the nature of everyday news' (p. 197). Thus, by contrast with the radical role, the collaborative role typically implies a cooperative relationship with established authority, one that is often practised (for example, in emergencies or wartime) but rarely admitted in conventional accounts of the 'free press'.

We argue that climate crisis necessitates more facilitative and radical journalism. Even at their best, corporate media do not stray for long beyond the monitorial/reportorial function into the terrain of consciously facilitating public engagement, let alone providing counter-hegemonic narratives or supporting popular mobilization.

Christians *et al.* summarize the space that each role occupies in the political spectrum (pp. 181–82). They suggest that the radical role corresponds with 'revolutionary' ideology, seeing journalism as 'an instrument for challenging and changing political and economic systems'. The monitorial and facilitative roles represent the reformist ideology of improving the system, whereas the collaborative role represents conservative ideology: the media as active instruments for preserving the system. Different alignments are quite possible, however. The radical role could stimulate necessary reforms without overthrowing the system. The monitorial role, if it means reporting primarily on the statements of the powerful or demonizing threats to the system, could be quite conservative. And the collaborative role could see journalists supporting revolutionary governments (such as Venezuela under Hugo Chavez) pursuing fundamental change. Such historical contingencies do not alter the key point: journalism is an inherently political practice, and climate crisis forces a rethinking of its fundamental public purposes. We turn now to how our book addresses this question.

Organization of the book

We have endeavoured throughout this work to acknowledge the 'crisis' element of climate change, and to evaluate the current challenges and potential opportunities that reporting climate crisis presents for journalism. Chapter One explores the relationship among journalism, democracy as its traditional touchstone in the West and the emergent climate crisis. What are relevant normative benchmarks for evaluating journalism's public performance in this new context? Several broad models of democracy and media – market liberal/competitive elitist, public sphere liberalism and radical democracies – are briefly outlined. Does climate crisis disrupt these models? How do major possible political responses to climate crisis, such as climate justice and the rights of nature, challenge conventional rationales for liberal democracy and embellish journalism's four democratic roles?

Chapter Two critically reviews the 'information deficit' approach to communication, and discusses key barriers preventing deeper public engagement with climate change. Drawing from research in environmental communication (and related disciplines), it explores four strategies for improving climate journalism: prioritize audiences most likely to engage with climate news as an 'issue public' (rather than using a 'one-size-fits-all' approach addressed to a mass audience); make greater use of a politically and ethically oriented climate justice frame; foreground and 'activate' intrinsic, biospheric cultural values which are most strongly correlated with pro-environmental subjectivity; and cultivate social norms of civic engagement and political efficacy with greater attention to the stories, experiences and emotions of people and communities working together to address climate change. While these strategies pose challenges to conventional reporting practices, they keep faith with journalism's basic democratic mandate to facilitate active public engagement with the most pressing contemporary issues, and hold promise to renovate and invigorate climate journalism.

Chapter Three selectively overviews the media strategies of environmental groups in recent history, particularly the movement's successful use of 'media stunts'. These have become a staple of campaigns by contemporary political movements and their media. In the context of discussions about radical media, social movement media, and the inexorable connections between the two in successful political action and social change, this chapter argues that dominant media coverage of environmental conflict and environmentalism is more likely to reflect the dispassionate 'monitorial' role of the media, and in some cases the 'facilitative' role, but rarely the radical role. Although most environmental conflict involves political contestation between industry, government and environmental lobby groups and activists, conventional media do not often amplify the views of the latter. That potentially radical role is more often undertaken by alternative media and their journalists who have better institutional supports for it.

How, then, can frames more productive to public engagement with climate politics be nourished within the media system? Chapter Four considers three levels of potential change. First, best practices are schematically categorized. Second, it considers two challenger paradigms (integrated editorial philosophies and methods) that have emerged since the 1990s – Civic Journalism and Peace Journalism. These hold promise for promoting public engagement, journalism reflexivity and story contextualization, but have faced limits of implementation within conventional news organizations. A third option entails incubating Climate Crisis Journalism within different structures – alternative media – whose currently marginal status in the political economy of the media provides an additional rationale for the reform of the policy framework of media industries.

A considerable body of scholarship has documented the many shortcomings of commercial media coverage of climate change, including its failure to offer citizens meaningful avenues to participate in climate solutions. Much less attention, however, has been devoted to emerging communication and journalism practices intended to motivate deeper public engagement with climate politics. Chapter Five begins to address this gap through interviews with advocates and journalists in British Columbia who have played leading roles in a thriving, evolving independent local media ecosystem which offers alternative perspectives on democracy, ecology and citizenship. This chapter focuses on the divergent responses of participants to frames of political conflict and polarization which often dominate media coverage and public discourse in the province. For some, these frames alienate the public from civic engagement, necessitating a corrective emphasis upon pragmatic, bipartisan solutions which emerge when cooperation, compromise and dialogue displace stories of conflict. Others, however, insist that such stories – especially in a region where the fight over neoliberal extractivism has become so pronounced – are both inescapable and represent a valuable opportunity to get citizens involved in political struggle.

Previous scholarly work has found a particular bias in mainstream Australian news media reporting of climate change, fostering scepticism in the public sphere. But the performance of Australia's alternative and independent media in communicating climate change has been largely neglected. Chapter Six addresses this gap, presenting new research that focuses on independent and alternative media reporting of a significant moment in climate change policy and action – the U.N. Conference of Parties Paris talks in 2015. It analyzes several key ways in which such media are reporting climate change in manners often the antithesis of mainstream reporting – with an absence of climate scepticism, the inclusion of voices usually marginalized, and an overall ambition to critique the hegemonic structures that support fossil fuel development.

Chapter Seven likewise presents new empirical data, and considers the impact of ongoing changes in the news media landscape on the ability of media adequately to report the 'big issues' of our time. Specifically, it considers both the limitations and new opportunities that the digitization of news content presents for quality coverage of complex social and political issues, such as environmental crises, drawing upon a case study of *The Guardian*'s 'Keep it in the Ground' campaign. The chapter questions what space now exists in the media system for climate justice and accurate representation of environmental issues. What discernible impact has the plethora of online news sites, social media platforms and discussion boards had on public information and public action? Have online journalism and the digital revolution in news content provided climate justice communication with a more secure future? What do the contributions of alternative media and alternative journalism suggest about the 'wisest' way we might deliver journalism about, and at a time of, climate crisis?

Our conclusions draw together the core themes of these chapters, finding the relevance of both a facilitative and a radical role for journalism and arguing for support of the sites in which these forms might best succeed. The inadequacy of conventional news media's climate journalism is evidence of their 'market failure'. Supporters of more effective and empowering climate communication have potential common cause with emergent movements for democratic media reform.

References

Almiron, N. (2010) *Journalism in Crisis: Corporate Media and Financialization* (translated by W. McGrath), Cresskill, NJ: Hampton Press.

Anderson, A. (2009) 'Media, politics and climate change: Towards a new research agenda', *Sociology Compass* 3: 166–82. Accessed at http://onlinelibrary.wiley.com/doi/10.1111/j.1751–9020.2008.00188.x/abstract.

Atton, C. (2002) *Alternative Media*, London: Sage.

Atton, C. and Hamilton, J. (2008) *Alternative Journalism*, London: Sage.

Beck, U. (1992) *Risk Society: Towards a New Modernity*, London: Sage.

Bennett, W.L. (2009) *News: The Politics of Illusion*, 8th edn, New York: Pearson/Longman.

Benson, R., Blach-Orsten, M., Powers, M. and Willig, I. (2012) 'Media systems online and off: Comparing the form of news in the United States, Denmark, and France', *Journal of Communication* 62: 21–38.

Boyce, T. and Lewis, J. (2009) 'Climate change and the media: The scale of the challenge', in T. Boyce and J. Lewis (eds), *Climate Change and the Media*, New York: Peter Lang, pp. 3–16.

Boyd-Barrett, O. (2011) 'Review of *Journalism in Crisis: Corporate Media and Financialization* by N. Almiron', *Journalism & Mass Communication Quarterly* 88(2) (Summer 2011): 449–50.

Boykoff, M.T. and Boykoff, J.M. (2004) 'Balance as bias: Global warming and the US prestige press', *Global Environmental Change* 14: 125–36.

Boykoff, M.T. and Yulsman, T. (2013) 'Political economy, media, and climate change: Sinews of modern life', *WIREs: Climate Change* 4(5): 359–71. Accessed at http://doi.org/10.1002/wcc.233 and http://proxy.lib.sfu.ca/login?url=http://search.ebscohost.com/login.aspx?direct=true&db=eih&AN=89769070&site=ehost-live.

Christians, C., Glasser, T., McQuail, D., Nordenstreng, K. and White, R.A. (2009) *Normative Theories of the Media: Journalism in Democratic Societies*, Urbana and Chicago: University of Illinois Press.

Climate Change Media Partnership (2011) 'Why the media matter in a warming world: A guide for policymakers in the global south', *Policy Brief*. Accessed at http://connect4climate.org/images/uploads/resources/CCMiPpaper.pdf.

Collins, S. and Rose, J. (2004) 'City voice: An alternative to the corporate model', *Pacific Journalism Review* 10(2): 32–45.

Cottle, S. (2009a) *Global Crisis Reporting: Journalism in the Global Age*, Maidenhead, UK: Open University Press/McGraw Hill.

—— (2009b) 'Series editor's preface: Global crises and the media', in T. Boyce and J. Lewis (eds), *Climate Change and the Media*, New York: Peter Lang, pp. vii–xi.

Curran, J. (2000) 'Rethinking media and democracy', in J. Curran and M. Gurevitch (eds), *Mass Media and Society*, 3rd edn, London: Arnold, pp. 120–54.

—— (2002) *Media and Power*, London and New York: Routledge.

Dann, E.G. (2016) 'Talking about wildfires and climate change isn't playing politics', *The Huffington Post* (May 31). Accessed at http://www.huffingtonpost.ca/g-elijah-dann/fort-mcmurray-fire_b_9890178.html.

De Souza, M. (2016) 'Justin Trudeau criticizes Elizabeth May's Fort McMurray climate connection', *National Observer* (May 4). Accessed at http://www.nationalobserver.com/2016/05/04/news/fort-mcmurray-fires-related-global-climate-crisis-says-elizabeth-may.

Donsbach, W. and Klett, B. (1993) 'Subjective objectivity: How journalists in four countries define a key term of their profession', *Gazette* 51(1): 53–83.

Downing, J.D.H., with Ford, T.V., Gil, G. and Stein, L. (2001) *Radical Media: Rebellious Communication and Social Movements*, 2nd edn, Thousand Oaks, CA: Sage.

Forde, S. (1997) 'A descriptive look at the public role of the Australian independent alternative press', *AsiaPacific Media Educator* 3: 118–30.

—— (2011) *Challenging the News: The Journalism of Alternative and Community Media*, Basingstoke, UK: Palgrave Macmillan.

Foxwell-Norton, K. (2015) 'Community and alternative media: Prospects for 21st century environmental issues', in C. Atton (ed), *The Routledge Companion to Community and Alternative Media*, London: Routledge, pp. 389–99.

Franklin, B. (2012) 'The future of journalism', *Journalism Practice* 6: 595–613.

Freedman, D. (2010) 'The political economy of the "new" news environment', in N. Fenton (ed), *New Media, Old News: Journalism & Democracy in the Digital Age*, Los Angeles: Sage, pp. 35–50.

Fuchs, C. and Sandoval, M. (2015) 'The political economy of capitalist and alternative social media', in C. Atton (ed), *The Routledge Companion to Alternative and Community Media*, New York: Routledge, pp. 165–75.

Gunster, S. (2011) 'Covering Copenhagen: Climate politics in B.C. media', *Canadian Journal of Communication* 36(3): 477–502.

—— (2012) 'Radical optimism: Expanding visions of climate politics in alternative media', in A. Carvalho and T.R. Peterson (eds), *Climate Change Politics: Communication and Public Engagement*, Amherst, NY: Cambria Press, pp. 239–67.

Haas, T. (2004) 'Alternative media, public journalism and the pursuit of democratization', *Journalism Studies* 5(1): 115–21.

Haas, T. and Steiner, L. (2006) 'Public journalism: A reply to critics', *Journalism* 7(2): 238–54.

Hackett, R.A. (2008) 'Objectivity in reporting', in W. Donsbach (ed), *The International Encyclopedia of Communication*, Malden, MA: Blackwell Publishing. Accessed at http://www.communicationencyclopedia.com.proxy.lib.sfu.ca/subscriber/book.html?id=g9781405131995_9781405131995.

Hackett, R.A. and Carroll, W.K. (2006) *Remaking Media: The Struggle to Democratize Public Communication*, New York and London: Routledge.

Hackett, R.A. and Zhao, Y. (1998) *Sustaining Democracy? Journalism and the Politics of Objectivity*, Toronto: Garamond [now University of Toronto Press].

Hadl, G. and Dongwon, J. (2008) 'New approaches to our media: General challenges and the Korean case', in M. Pajnik and J.D.H. Downing (eds), *Alternative Media and the Politics of Resistance: Perspectives and Challenges*, Ljubljana: Peace Institute, pp. 88–110.

Hallin, D. and Mancini, P. (2004) *Comparing Media Systems*, Cambridge, UK: Cambridge University Press.

Harcup, T. (2013) *Alternative Journalism, Alternative Voices*, London and New York: Routledge.

Hopke, J.E. (2012) 'Water gives life: Framing an environmental justice movement in the mainstream and alternative Salvadoran press', *Environmental Communication: A Journal of Nature and Culture* 6(3): 365–82.

Hughes, L. and Steffen, W. (2013) 'Be prepared: Climate change and the Australian bushfire threat', in *Climate Council of Australia*, Sydney. Accessed at http://www.climate-council.org.au/uploads/c597d19c0ab18366cfbf7b9f6235ef7c.pdf.

Hutchins, B. and Lester, L. (2015) 'Theorizing the enactment of mediatized environmental conflict', *International Communication Gazette* 77(4): 337–58.

Ireland, J. (2013) 'UN official "talking through her hat" on bushfires and climate change, says Tony Abbott', *The Sydney Morning Herald* (October 23). Accessed at http://www.smh.com.au/federal-politics/political-news/un-official-talking-through-her-hat-on-bushfires-and-climate-change-says-tony-abbott-20131023-2w0mz.html.

Jankowski, N. (2003) 'Community media research: A quest for theoretically-grounded models', *Javnost* 10(1): 1–9.

Kenix, L.J. (2008) 'Framing science: Climate change in the mainstream and alternative news of New Zealand', *Political Science* 60(1): 117–32.

——— (2011) *Alternative and Mainstream Media: The Converging Spectrum*, London and New York: Bloomsbury Academic.

Klein, N. (2014) *This Changes Everything: Capitalism vs the Climate*, Toronto: Knopf Canada.

Lynch, J. and McGoldrick, A. (2005) *Peace Journalism*, Stroud, UK: Hawthorn.

Maras, S. (2013) *Objectivity in Journalism*, Cambridge, UK: Polity.

McChesney, R.W. (2003) 'The problem of journalism: A political economic contribution to an explanation of the crisis in contemporary US journalism', *Journalism Studies* 4(3): 299–329.

McChesney, R.W. and Nichols, J. (2010) *The Death and Life of American Journalism*, Philadelphia: Nation Books.

Oreskes, N. and Conway, E. (2010) *Merchants of Doubt: How a Handful of Scientists Obscured the Truth on Issues from Tobacco Smoke to Global Warming*, New York: Bloomsbury Press.

Raymond Williams. (1989) *Resources of Hope: Culture, Democracy, Socialism*, New York and London: Verso.

Rodriguez, C. (2001) *Fissures in the Mediascape: An International Study of Citizen's Media*, Cresskill, NJ: Hampton Press.

Siebert, F., Peterson, T. and Schramm, W. (1956) *Four Theories of the Press*, Urbana: University of Illinois Press.

Slezak, M. (2016a) 'Scientists resort to advertising to get Great Barrier Reef crisis in Queensland paper', *The Guardian* (April 16). Accessed at http://www.theguardian.com/environment/2016/apr/21/scientists-resort-to-advertising-to-get-great-barrier-reef-crisis-in-queensland-paper.

———— (2016b) 'Great Barrier Reef bleaching made 175 times likelier by human-caused climate change, say scientists', *The Guardian* (April 28). Accessed at http://www.the-guardian.com/environment/2016/apr/29/great-barrier-reef-bleaching-made-175-times-likelier-by-human-caused-climate-change-say-scientists.

Smeltzer, S. (2008) 'Biotechnology, the environment, and alternative media in Malaysia', *Canadian Journal of Communication* 33(1): 5–20.

Upton, J. (2015) 'Media contributing to "hope gap" on climate change', *Climate Central* (March 28). Accessed at http://www.climatecentral.org/news/media-hope-gap-on-climate-change-18822 on April 28, 2015.

Zelizer, B. (2011) 'Journalism in the service of communication', *Journal of Communication* 61(1): 1–21.

Chapter 1

Democracy, climate crisis and journalism

Normative touchstones

Robert A. Hackett

Climate change is not just a scientific, technical or economic matter. It poses profoundly *ethical* and *political* challenges to human institutions, including journalism.

This chapter explores some normative dimensions of a triangular relationship – journalism, democracy and climate crisis – and raises questions about whether global climate crisis necessitates different ethical touchstones for assessing and potentially guiding journalism's practices, structures and texts.

In Western debates about journalism ethics, journalism is evaluated above all by how well it makes democracy work. But that statement needs immediate qualification. Democratic purposes do not exhaust journalists' professional self-definition. Globally, journalism takes many forms, accepts different ethos and experiences different relationships with economic and political systems, even within the advanced capitalist states (*cf.* Hallin and Mancini 2004); it 'does have a life outside democracies' (Josephi 2013).

Nor is democracy a goal universally shared. Concerns for security, stability, social harmony, law and order and economic well-being often trump commitment to democracy. In the context of economic stagnation, superpower nostalgia and political turbulence, a majority of Russians apparently feel that strong leadership is more important than democracy (Pew Research Center 2012). In Singapore, paternalistic governments proclaim a commitment to 'Asian values'. In capitalism's heartland, for millions of non-European and Indigenous people, American democracy has historically meant expropriation, slavery, even genocide (Hackett and Carroll 2006: 11). Nor is democracy necessarily liberal, in the sense of protecting individual and minority rights. In recent history, various countries have elected governments that legalize discrimination against religious or ethnic minorities. Moreover, as the late Canadian political theorist C.B. Macpherson (1966) argued, the 'real world of democracy' can encompass, in theory at least, vastly different political and economic systems – from free market liberalism, to the Marxist project of building a classless society, and the postcolonial development-oriented states of the global South.

Yet democracy, in the sense of legitimacy derived from a popular mandate, is undeniably the globally dominant form of political legitimation: 'political regimes of all kinds describe themselves as democracies', argues a leading political theorist, David Held (2006: 1). As the rebellious peasant haranguing King Arthur in *Monty Python and the Holy Grail*, actor Michael Palin expresses the point comically: 'supreme executive power derives from a mandate from the masses, not from some farcical aquatic ceremony.'

But there is considerable variation within political theory on three key issues: What is the core meaning of democracy? Is it necessarily a good thing, and why? What are its actual practices, boundaries and preconditions (Cunningham *et al.* 2015)? Different models of democracy have different answers, and accordingly different expectations (typically only implicit) for the legitimate role of communication media. As discussed in the Introduction to this book, Christians *et al.* (2009) have identified several different key roles that democratic journalism is expected to perform: the *monitorial* function of reporting on publicly relevant events and developments, the *facilitative* role of nourishing democratic public spheres and popular engagement with public issues, the *radical* role of exposing injustice and encouraging social change, and the *collaborative* role of supporting broader social purposes and institutions.

These benchmarks have differential relevance to contending models of democracy, to which we now turn. Read with caution: no definitive review is attempted here! To highlight their divergences, we override nuance and massage models of democracy into a trio: market liberalism, deliberative democracy and radical egalitarian democracy. Each of these implies (but does not rigidly entail) different expectations of how journalism should function, what its ethical principles and practices should be, and what kind of institutional and legal frameworks best support it.

Democracies and their journalisms[1]

For much of the Cold War period, a leading tradition in the West for understanding democracy and the media was liberal-pluralism. It sees democracy as 'a process of competition between diverse interests and multiple power centres' (Curran 2011: 80); like politics, the media marketplace should be open to such competition without being restrained by either government policy or the requirements of objectivity. Instead, it embraced advocacy and partisanship and a 'free-for-all market approach to journalism' (*ibid*). While it still provides the underlying normative rationale for the British approach to print journalism, it has arguably been supplanted – at least in elite policy discourse – by approaches informed by neoliberalism, with its assumptions of atomized individuals making rational, self-interested choices in economic and political marketplaces.

Market liberalism and elitist democracy

Since the 1980s, the 'free market' vision of democracy has gained political and cultural hegemony in the U.S. and the U.K. Democracy is seen not as an end in itself, but as normally the best institutional arrangement to maintain political stability and a liberal political culture characterized by individual rights and choice, particularly economic rights of ownership, contract and exchange. 'The market' is seen as the best organizing principle for not only the economy, but also society more generally; it is taken to be the realm of 'freedom' and individual choice, by contrast with politics as a necessary evil, the realm of coercion.

Free market politicians often adopt populist rhetoric, bashing 'liberal' cultural elites. But the same politicians – such as Republicans in the U.S., Conservatives in Canada, the inappropriately named (right-wing) 'Liberal Party' in Australia and so on – also introduce voter suppression legislation that makes it more difficult for socially disadvantaged groups to exercise their franchise. This is no coincidence, as the 'free market' preference for minimal government – apart from maintaining social order through the State's military and police powers – actually fits well with a 'competitive elitist' version of democracy classically articulated by the Austrian-American economist Joseph Schumpeter (1976; cited in Baker 2002: 130; see also Held 2006: 126–57). In this view, the complexity of modern political issues, the vulnerability of the masses to irrational and emotional appeals, and the risk of overloading the political system with competing demands makes ongoing public participation neither necessary nor even desirable. Democracy is seen as a procedure for selecting between 'competing teams of elites' (Curran 2011: 80), with citizen participation confined mainly to voting every few years – essentially, the role of consumers in a political marketplace. Policy-makers can be held sufficiently accountable through periodic elections, the entrenchment of individual political rights and a free press.

For its part, the press 'need not provide for nor promote people's intelligent political involvement or reflection', since 'meaningful understanding of social forces and structural problems is beyond the populace's capacity' (Baker 2002: 133); nor need it raise fundamental questions about State policy or the social order. So much for the public sphere.

Nevertheless, journalism does have positive monitorial roles in the market liberal model. By exposing corruption and the abuse of power, the press should act as a watchdog on government, which is considered the main threat to individual freedom. And journalism, particularly the 'quality' press, should report intra-elite debates and circulate 'objective' information useful to elites themselves – a mandate for journalism articulated almost a century ago by the legendary American political columnist Walter Lippmann (1922).

Still, the objectivity ethos should not be imposed on the press. Media owners are free to flog their own views, held in check by consumers' presumed ability to punish excessively biased or inaccurate media in the marketplace. Or not. Market-driven media do not necessarily generate a self-correcting marketplace of ideas. In the U.S., a combination of broadcasting deregulation, political polarization and the explosion of online outlets has enabled citizens to live in media cocoons that nurture their own prejudices. While Rupert Murdoch's Fox News network famously led the way in creating a self-validating political universe indifferent to fact-checking or objectivity, it is not unique.

Public sphere liberalism/deliberative democracy

The elitist model of democracy has been criticized on many grounds. Its negative view of citizens' participation is unduly pessimistic. In referenda and elections on fundamental issues, citizens have sometimes shown a remarkable capacity for learning and civic engagement. Conversely, scandals such as the apparent manipulation of security intelligence by the U.S. and U.K. governments before they invaded Iraq in 2003 suggest that the elitist model overestimates the competence and accountability of policy-makers without ongoing public participation.

Similarly, the related market liberal approach to democracy overlooks the excessive power of concentrated wealth in policy-making processes. It dismisses the threat to political equality and even meaningful individual freedom posed by the growing gap between rich and poor, a gap reinforced by neoliberal policies of cutbacks to social programs, public services and taxation of the wealthy. It ignores the erosion, by a culture of acquisitive individualism, of the sense of community underpinning democratic governance. And from an ecological perspective, market liberalism's adulation of property rights and the pursuit of private gain sits uneasily beside the green acceptance of collective solutions and governmental intervention to environmental challenges, and the need for constraints on individual consumption (Martell, cited in Carter 2007: 68). Market liberalism is even less likely to entertain radical green challenges to capitalism as an inherently ecologically destructive system driven by the constant expansion of capital (Klein 2014; Magdoff and Foster 2011).

Such considerations have strengthened an alternative vision that accepts the elitist democrats' support for individual rights and an independent 'watchdog' press, but places a much higher value on popular participation through established political channels. Participation can be valued as a means both to produce more just and legitimate policies, and to develop the democratic capacities of citizens. Participation is not simply a question of voting, but of engagement in deliberation, understood as:

> the exchange of reasons under conditions of fairness and equality among citizens who are open to competing arguments and, where necessary, accommodating alternative views. In this sense deliberative democracy takes seriously the idea that preferences are formed as part of the political process.
>
> (Niemeyer 2013: 430)

Deliberative democracy would also be *inclusive*, in that people who are affected by a decision 'have the opportunity to deliberate and provide input into the decision-making process', and all issues of interest to civil society are addressed, including those related to the environment (pp. 430, 433).

In strong versions of such a participatory and deliberative democracy, not only public opinion but also government policy is shaped through civil society's collective deliberations about its future. Deliberation is *consequential*, in that citizens' deliberations would be reflected in the decision being made (p. 430). The public sphere is culturally and institutionally central in this model.

What, then, are journalism's key tasks in this model? They include the monitorial role and especially the facilitative role of nourishing the public sphere by encouraging public participation, providing a civic forum to sustain both pluralistic political competition and the search for social consensus, and stimulating general interest, public learning and civic engagement vis-à-vis the political process (Norris 2000: 25–35). To sustain deliberative democracy in a pluralistic society, Baker (2002: 129–53) advocates two offsetting types of news media: a segmented system that provides each significant cultural and political group with a forum to articulate and develop its interests; and public service media that can facilitate the search for society-wide political consensus by being universally accessible, inclusive (civil, objective, balanced and comprehensive), and thoughtfully discursive, not simply factual. Given the role of segmented media to represent particular and sometimes oppositional groups, objectivity is not a universal norm in journalism for deliberative democracy, but facilitating the formation of public opinion certainly is.

Radical democracies

The deliberative democrats' critique is arguably an advance over market liberal/competitive elitist theory that justifies public political passivity and the attendant reproduction of inequality (Curran 2011: 81). Deliberative democracy poses a challenge to a supposedly representative democracy that is in practice dominated by political parties, lobbyists and professional image-makers. But its critique is blunted by a basic presupposition – the search for a reason-based consensus within the taken-for-granted hegemonic framework of contemporary capitalism. More radical visions of democracy problematize that assumption on the grounds that power imbalances can skew the process and outcomes of deliberation: 'the rhetoric of "being reasonable" can be deployed by the powerful to exclude what they regard as "unreasonable", while the pursuit of consensus can obscure irreconcilable conflicts of value and interest in a manipulative form of closure' (Curran 2011: 81).

The radical tradition reminds us that equality within deliberative venues – public spheres – is unlikely if inequality is rampant within the society at large. The models of democracy previously discussed developed, for the most part, in European nation-states and expanding American capitalism. Both of those contexts nourished but also constrained democracy. The French Revolution declared Liberty, Equality and Fraternity as core principles, but when the nation-state becomes the container and guarantor of political rights, what becomes of non-citizens within the national territory – say, African immigrants or West Asian refugees? Capitalism laid some of the cultural and legal building blocks of contemporary democracy, both positively, through (for example) generating the legal fiction of the freedom and equality of individuals as agents able to enter into contracts (for employment, investments, commodity purchases), and negatively, through creating the urban industrial working classes whose struggles against exploitation and for social and political rights put the 'democracy' in liberal democracy.

On the other hand, capitalism constrains the extent of popular sovereignty. Its economic inequalities are easily translated into political inequality. As Donald Trump's presidential campaign shows, billionaires do not need policy competence to acquire political influence; their wealth and celebrity status (under neoliberalism) as 'successful businessmen' (and television celebrities, in Trump's case) suffice. Moreover, capitalism imposes its own limits on policy options. When electoral majorities have supported policies that challenge its fundamentals, capital may be prepared to discard the democratic process, and revert to authoritarian or even fascist government. Some cases in point: Chile's military coup in 1973, ousting Salvador Allende's elected socialist government; the attempted coup against Venezuela's Hugo Chavez in 2002; and European Union pressure on governments such as Greece to enact draconian austerity measures without a popular mandate in 2015.

Capitalism and democracy have thus co-existed in uneasy tension for over a century; indeed, the containment of democracy has been a precondition for capitalism's survival. Market liberalism is not only an economic system but a political one; in the radical view, the state is not only liberal democratic (at best) but also capitalist.

Socialists start from an oppositional and holistic view of power in the contemporary social order, one that bell hooks (1981) labels 'white supremacist capitalist patriarchy' in order to emphasize the connectedness of distinct forms of oppression. Marxist theorists like Ralph Miliband (1973[1969]: 132) noted that the 'private control of concentrated industrial, commercial and financial resources' generated 'pervasive and permanent pressure' on the state, but the same can be said of the public sphere. The voices of those with the greatest economic and social capital can overwhelm the rest – whether through the ability to mount expensive propaganda and lobbying campaigns, the ideological capacity to win consent to their definitions of social reality, and/or the uneven distribution of the forms of cultural competence (such as written and digital literacy) that are deemed necessary to gain standing in public discourse. This is partly what is at stake in Antonio Gramsci's (1971) conception of hegemony – the capacity of a dominant coalition or 'power bloc' to maintain its rule through cultural leadership, not simply coercion. It is also a question of the marginalization of gendered, racialized and other subordinated collectivities. Indeed, what is at stake in the formation of publics and contestation within them is not only particular policy options, but the forms of cultural capital that are to be valued. In the context of contestation over energy projects, the weight given to aboriginal peoples' oral traditions and knowledges (as distinct from European settlers' documents and currencies, for example) matters a great deal.

Interestingly, feminist and democratic theorist Iris Marion Young (2001) used the example of greenhouse gas policy to illustrate how such power can shape discourse. Conventional debates focus on questions such as by how much emissions should be reduced, and how to distribute such reductions between richer industrialized countries and the less developed ones. Such debates assume that the economies of all developed countries must rely on fossil fuels to sustain their accustomed lifestyles and rule off the agenda visions of a world no longer economically reliant on carbon emissions (p. 687).

In a socialist view of power and democracy, what political roles are expected of news media? They would endorse the watchdog and public sphere functions, but they would celebrate more active advocacy of social change. Journalism should actively counterbalance power inequalities generated elsewhere in society. Expand the scope of public awareness and political choice by reporting events and voices which are socially important but are outside, or even opposed to, the agendas of elites. Enable horizontal, intersectional communication between subordinate groups, such as workers, women, ethnic minorities, and social movements as agents of democratic renewal (Hackett and Carroll 2006, Chapter 3). Engage in open advocacy of anti-capitalist politics. An example of such intersectional advocacy communication is the magazine *Canadian Dimension*, which through its pages, website and editorial board, brings together voices from different progressive social movements; its motto is 'for people who want to change the world'. By giving public voice to civil society, media can facilitate needed social change, power diffusion, and popular mobilization against social injustices.

Not surprisingly, this radical role is more likely to be expressed in 'alternative' media outlets that have positive relationships with movements for social change than in 'mainstream', corporate or state-owned media more closely tied to established power. Alternative media are discussed later in this book, but at this point we can ask whether such media would strive for objectivity. There is no simple answer, and it partly depends on how we understand objectivity – as a stance of neutrality between contending sides, or as a search for truth that could lead a reporter to come down on one side rather than another. To the extent that they are committed to progressive social change, journalists in radical media would favour advocacy over a stance of neutrality. As David Barsamian, the Colorado-based producer of Alternative Radio put it, 'I reject objectivity because I don't want to give a voice to injustice' (Hackett and Gurleyen 2015: 56). Extensive interviews with alternative journalists in the U.S., U.K. and Australia reveal a widespread rejection of neutrality and detachment in favour of open advocacy (Forde 2011: 114). Most radical journalists would reject mainstream media's claim to providing objective accounts of the world, but only some would go so far as to reject the very possibility of objectivity as inherently fraudulent and deceptive, a smokescreen for particular ideologies. Other alternative journalists might well see themselves as actually practising a better version of objectivity – by speaking truth to power and drawing the implications of a valid understanding of the world, with its injustices, for political change.

A 'post-Marxist' turn

Ernesto Laclau and Chantal Mouffe (1985) inaugurated a now-influential offshoot from the socialist tradition – radical democracy – which has challenged classical Marxism's focus on class inequality and economic exploitation as the primary axis of domination. Radical democrats have abandoned the dream of a fixed and classless utopia, as new forms of exclusion and oppression will constantly arise in human society. Radical democracy is 'post-Marxist' in rejecting the reduction of ideology to class interests, and the essentializing of people's identities to a single category (like 'the worker') based on those interests; identities are seen as contingent and constructed through discourse. It is nevertheless 'radical' in encompassing 'the extension of democracy into ever-widening areas of the social' (Carpentier and Cammaerts 2006). And radical democracy has helped to identify a blind spot of deliberative democracy: it both overvalues consensus and overestimates the possibility (and necessity) of achieving it – particularly on an issue as complex and fraught with contending interests as climate crisis. Instead, dissent and confrontation are to be valued as agents of both inclusion and systemic change and as prophylactics against identity-based politics. Mouffe (2002) argues that 'too much emphasis on consensus, together with aversion towards confrontations, leads to apathy and to disaffection with political participation.' She offers a model of 'agonistic pluralism' that 'acknowledges the role of power relations in society and the ever present possibility of antagonism'. In her view:

the aim of democratic institutions is not to establish a rational consensus in the public sphere but to defuse the potential hostility that exists in human societies by providing the possibility for antagonisms to be transformed into agonism. What I mean by this is that in democratic societies the conflict cannot and should not be eradicated but that it should not take the form of a struggle between enemies (antagonism) but between adversaries (agonism).

(p. 48)

In this model, what is the expected role of journalism? Mouffe has addressed this question directly. Media should 'contribute to the creation of agonistic public spaces in which there is the possibility for dissensus to be expressed or different alternatives to be put forward' (Carpentier and Cammaerts 2006: 974). Journalism should be objective regarding straightforward factual truths, but pluralistic with respect to interpretations, aspects and perspectives. Pluralism *through* the media – enabling various groups to participate in public dialogue – needs to be paralleled by pluralism *of* the media – diversity of ownership and broadened participation in media production and decision-making.

Agonistic pluralism implies actively seeking dissenting views, and allowing space for passionate disagreement, while containing debate within the bounds of democratic co-existence. Journalistic practices, respectful of diversity and openness:

need to be built on a balanced approach between dialogue/deliberation and debate, between (information regarding) social consensus and social conflict, and between (information about) solutions and problems.

(Carpentier and Cammaerts 2006: 972)

Journalism should thus 'offer a counterweight for exclusionary hegemonic processes that restrict the access of discourses and identities to the media system', which in turn requires 'sensitivity for the conflict-ridden nature of the social and the political . . . and the workings of ideology and hegemony' (Carpentier and Cammaerts 2006: 974). We might say that Mouffe's model hybridizes the facilitative and radical roles of journalism – providing a forum for broad dialogue with particular concern for empowering dissident citizens of the status quo. Her work provides a rationale for excluding political forces that are beyond the 'frontier' of legitimate democratic debate, such as overtly Nazi or extreme Islamicist parties (Hansen and Sonnichsen 2014: 269).

In short then, each model of democracy respectively emphasizes three roles for journalism with potentially positive contributions for addressing climate crisis: monitoring the state's exercise of power, facilitating public engagement and discussion, and mobilizing movements for change against an environmentally destructive status quo. But in the context of climate change and other crises, is democracy itself still relevant?

Does climate crisis disrupt these models of democracy and journalism?

A tsunami of multiple and intersecting crises (Cottle 2009: 1–25) – climate change and resource depletion; actual and potential political violence, exemplified by a decade of 'terror war' (Kellner 2003); stateless refugees and economic migration, fleeing the consequences of political violence, poverty and climate devastation; financial instability and periodic meltdowns; globalized inequality and poverty – may comprise a state of 'planetary emergency', in the words of James Hansen, who is arguably America's best-known climate scientist. The existential threat of climate change in particular poses the need for extraordinarily rapid and widespread policy response. Can conventional political and economic institutions – notionally representative liberal democracy and market liberalism – react with sufficient urgency or even in an ecologically sustainable direction?

Besides straining institutions' adaptive capacities to the limit, climate crisis also calls into question conventional liberal rationales for democracy – their anthropocentrism. As we have argued, the classic theories of democracy stemming from the European Enlightenment differ in important respects, but they have in common the value of expanding individual autonomy and freedom, arguably premised upon not only the political defeat of tyranny but the spread of material prosperity. They take for granted the separation of humans from nature and the ethical priority of the former. It is a dichotomy challenged by emergent strands of green political thought (Carter 2007, Chapter 3).

The condition of planetary ecological emergency, however, forces a re-think. What if we have reached the 'end of growth'? Instead of the Enlightenment's exaltation of human development, suppose our starting point is instead the inherent embeddedness of human civilization in the broader ecosphere – and the realization that our abuse of that relationship is jeopardizing the continuation of the human project. New criteria and questions are then raised. How does political decision-making incorporate accountability to future generations and to non-human species? What kinds of restrictions do majorities need to place upon their own freedom now, in order to facilitate future freedoms for others? What does a politics of sustainability look like?

Eco-authoritarianism?

For some commentators, a politics of sustainability in the context of global crisis justifies declaring a state of emergency. James Lovelock (2010), the formulator of the Gaia hypothesis – the planet as a self-regulating single complex system, argued in a *Guardian* interview that 'even the best democracies agree that when a major war approaches, democracy must be put on hold for the time being. . . . Climate change may be an issue as severe as a war. It may be necessary to put democracy on hold for a while.'

If climate crisis threatens to overwhelm global society, it is tempting to think that we need top-down command-style messaging, backed by authoritarian policies, to bang reluctant heads together and force through the necessary remedies. Without going that far, certain professional analysts would, from ecological motives, intensify the elitism of the competitive elitist model discussed previously. Sociologist Anthony Giddens (2009) would insulate climate change decision-making from the regular democratic process, and others would place decision-making entirely in the hands of enlightened scientists (Shearman and Smith 2007; cited in Niemeyer 2013: 432). Still others pay tribute, however reluctantly, to the alleged efficiency of Asia's authoritarian regimes, particularly China, in achieving environmental and developmental goals (Beeson 2010).

To be sure, climate crisis poses exceptional challenges to liberal democracy. Its time-scale vastly exceeds the four-year election cycle, its global scope necessitates international cooperation beyond the national scope of the (notionally) democratic polity, and it may require immediate concrete sacrifices (like higher prices for fossil fuels) in order to obtain the uncertain and diffused benefits of reduced emissions. The worst impacts of global warming may be perceived as too distant in time or place, and in any case, too remote or difficult to be tackled by specific government policies. Small wonder that climate change, to date, has not been high on the agenda of electorates or governments in liberal democracies. David Held concludes that environmental thus challenges, including managing the global commons, create 'some of the most fundamental pressures on the efficacy of the nation-state and state-centric democratic politics' (Held 2006: 302). Eco-elitists are surely right on at least one point: governments need to address climate crisis *as* a crisis, with a much greater sense of urgency than they have to date. Given its agenda-setting role, now shared between conventional and online media, journalism could help convey that sense of urgency, through sustained reportage that connects the dots and explores options for solutions.

But such considerations do not mean that eco-elitism is an alternative preferable to democratic decision-making. Some eco-elitists dubiously assume that the world is already democratic, that we live in a 'cheeky world' where 'everyone can have their say' (Lovelock's phrase). On the contrary. The governing logics of most states in globalized capitalism are skewed towards capital accumulation, even apart from the influence of 'well-resourced interests'; without popular pressure, political elites are less rather than more likely to support remedial environmental policies (Niemeyer 2013: 432–3). Under existing relations of power and the global hegemony of neoliberal capitalism, planetary emergency law would yield regimes of plunder more naked and brutal than we already have, generating even faster ecological decline unchecked by the cries of its victims – the 'voices of the side effects' (Beck 1992: 61; cited in Cox 2010: 215). In moments of crisis, outright fascism remains a standby option for those who would 'install an authoritarian regime in order to preserve the main workings of the system' (Kovel 2007: 205). In the wake of increased migration and terrorist attacks, proto-fascist politicians and parties in the U.S. and Europe have spiked in popularity, and the once 'exceptional' national security measures of notionally democratic states are becoming normalized.

Greening capitalism?

Fortunately, most green political theorists and activists reject authoritarianism. They also tend to be sceptical of the currently dominant Western political model, competitive elitism and market liberalism, although some branches of the environmental movement, such as major conservationist organizations, tried to jump on the neoliberal bandwagon following the 1980 election of U.S. president Ronald Reagan (Klein 2014). The initial neoliberal response to climate change, which began to surface as a scientific and political issue in the 1980s, was to ignore or deny its reality (although Conservative British Prime Minister Margaret Thatcher briefly found it a useful rhetorical hammer against militant coal miners). When denial became increasingly untenable, with the important exception of the U.S. Republican Party, political and economic elites turned to appropriating and translating environmental concerns, integrating them into a process of technological and market-oriented modernization (Kenis and Lievens 2014: 543). To the extent that it is politically possible, the elites want to carry on business as usual, and/or use the climate crisis to generate extensions of market relations, such as cap-and-trade carbon emission schemes. We thus hear promises of technological fixes, from solar energy to carbon sequestration; of a green economy; and of change through individual consumers' voluntary actions like changing light bulbs and buying hybrid cars. The faith in technology and market mechanisms is paralleled by faith in elite-driven legislative reforms and negotiations, like the United Nations Climate Change Conference (COP21) in Paris in 2015.

It would be welcome news indeed if such reformism, promoted by articulate advocates like former U.S. vice-president Al Gore (2006), could effectively mitigate global warming. No need then to expend energy in mobilizing social movements, challenging economic structures, or transforming social relations. In such a scenario, democratic journalism might well highlight the collaborative role of supporting an elite-led transition towards greener capitalism, and lauding the alleged accomplishments of technology, markets and enlightened politicians. In many ways, that is how significant sections of the corporate press reacted to the outcome of the self-congratulatory Paris COP21 summit. It remains to be seen whether mainstream journalism will exercise the monitorial function of assessing corporate and governmental behaviour in relation to the COP21 targets.

But what if green capitalism is a chimera? What if a growth imperative is hard-wired into capitalism? What if a view of nature as a limitless cost-free gift to be exploited and contaminated as needed (for capital) puts market liberalism on a collision course with ecological sustainability? Greener, more efficient energy technology is clearly part of the solution, but without enforceable ceilings, it historically generates greater energy consumption rather than reduced greenhouse gas emissions. Technology to remove carbon from the atmosphere is so far little more than a fantasy. The actual proposals in the 2015 Paris climate summit are too weak to achieve the stated target of a 1.5 degree Celsius global warming ceiling and, furthermore, entail no real enforcement mechanisms (Dolack 2015). Such protocols are in any case trumped by 'free trade' agreements, like the Trans-Pacific Partnership, that are enforced with punitive sanctions and that override pro-climate policies like renewable energy subsidies, restrictions on dirty fuels, and 'buy local' rules (Klein 2014, Chapter 2). Voluntary action by individuals may be ethically commendable, and could be a springboard to participating in collective political action. On its own, however, it is nowhere adequate to the pace and scale of change needed, has an obvious 'free rider' problem, and may actually reinforce values and behaviours of unsustainable consumerism (Klein 2014: 212). Market-oriented, technological and voluntarist approaches may well divert energy away from the political solutions that are the most hopeful way forward (Machin 2013, Chapter 1). And in the political realm, the 'competitive elitist' model of democracy is losing its efficacy; it is alienating millions of citizens that it claims to represent, while magnifying the political power of anti-climate extractivist capital.

If the world needs alternatives that supplement or transcend green capitalism, then the facilitative and radical roles of journalism acquire added urgency.

From a liberal towards a radical public sphere?

Contrary to eco-authoritarianism and competitive elitism, the prospects for legitimate and effective climate change policy likely depend on more democracy, particularly the engagement, support and buy-in from the millions of people whose lives would be affected by it. In line with that assumption, and after earlier flirtations with authoritarianism and anarchism, mainstream political thought gravitated by the early 2000s towards deliberative democracy as the decision-making model most conducive to mobilizing collective action for dealing with climate crisis (Arias-Maldonado 2007; Carter 2007).

At the same time, the unfolding exigencies of environmental politics call into question conventional notions of the public sphere. Historically, in Habermas's narrative, public spheres developed in the context of shared (bourgeois) class interests (Calcutt and Hammond 2011: 152–3) and relatively culturally homogeneous European nation-states. Given the exclusion of workers (not to mention women), participants did not need to dispute and attempt to resolve fundamental conflicts of interest, and could thus potentially operate in a consensual, rule-bound way. How can this model of equal, rational deliberation be practised with respect to the issue of global warming, characterized by both shared and conflicting interests and values in a context of shrinking resources and growing risks? To use the metaphor of planet Earth as the Titanic, humanity has a shared interest in keeping the ship afloat but conflicting ideas on how to avoid the (melting) icebergs and competing claims to the few (illusory?) lifeboats, to which first-class passengers in the luxury suites of the economically developed countries already have privileged access. Meanwhile, the captain is forging full steam ahead, and the ship's survival may depend on replacing him.

In this admittedly stretched metaphor, the 'captain' may be difficult to identify, and the absence of an obvious 'target' has been a key challenge for the environmental movement (Kenis and Lievens 2014). Recent global activism, however, has nominated one: the fossil fuel sector – coal, gas, oil – whose business model depends on massive and largely unregulated greenhouse gas emissions. In his now-famous article, leading American environmentalist Bill McKibben (2012) argued that the main strategies of the environmental movement against global warming had failed. Big oil corporations have not been persuaded to switch to sustainable energy. Persuading individuals to change lifestyles takes far too long and is 'like trying to build a movement against yourself'. The political process (at least in the U.S., the world's largest greenhouse gas emitter) has yielded meagre results. The missing key, he argued, is to frame the fossil fuel industry as a reckless rogue, because rapid transformative change requires building a movement, and 'movements require enemies'.

Challenging and counterbalancing the power of Big Carbon has thus been adopted as a core strategy by groups such as 350.org, with its campaign for fossil fuel divestment. Besides transnational environmental movements, other struggles are joining the challenge – aboriginal peoples defending their traditional and treaty-entitled territory; farmers and ranchers defending their livelihoods; grassroots place-based communities resisting extractivist attacks on their water and land base; and religious leaders, now including Pope Francis, articulating the moral imperative for transformative action (Klein 2014). New coalitions among these groups constitute an emerging force for civilizational change. Their success requires horizontal communication between them, a weightier presence in national and global public spheres, and more effective participation in policy-making.

In that light, the concept of 'public sphere' needs to be conceptualized as operating within movements for social and ecological change, and *against* the political and economic sponsors of ecocide. This reading implies a rather expanded role for subaltern counter-public spheres (Fraser 1997), capable of developing and articulating the interests of dominated groups on a transnational basis, before re-engaging and challenging the keepers of the global order. If indeed the fossil fuel industry is the enemy of climate change solutions, then it is not appropriate to invite them into the solutions tent, at least not initially. Indeed, if powerful actors such as Exxon and BP remain hostile to effective climate change solutions, they may be beyond Mouffe's 'frontier' of legitimate democratic debate, noted previously. It may be necessary to work for their political elimination through divestment, break-up, or expropriation. Only the counter-public spheres of transnational social movements could make such a goal seem less fanciful.

Public spheres, then, are inherently conflictual. Effective climate change responses, particularly those that challenge entrenched power, will likely require the mobilization of millions of ordinary people who refuse to take *No* for an answer – in short, the creation of a mass movement. Much of the political energy for transforming to a greener economy derives from activism that has become labelled 'Blockadia' – local community-based but transnationally connected resistance to extractivist mega-projects, whether coal mining expansion in Australia, coal-fired plants in India or bitumen-transporting pipelines in Canada (Klein 2014). Conflict and confrontation are keys to the success of such movements for radical change. Movements need 'collective action frames' that identify certain conditions as grievous, identify sources responsible for those conditions and propose remedies (Klandermans 2001).

This view of social movements as oppositional counterpublics connects with the radical view of power discussed previously. A democratic public sphere cannot be insulated from power hierarchies embedded in State, economy, gender and race; so long as they exist, they will tend to undermine equality of voice in the public sphere. Radical political theory nudges us towards considering not just political processes but the substantive distribution of power and rewards throughout the economy and society and towards a robust view of democracy as not just a political system but a type of society that nourishes developmental power – everyone's equal right to 'the full development and use' of their capabilities (Downing 2001: 43–4; Macpherson 1977: 114).

But if genuine democracy requires radical social change, the converse is also true. In the struggle for greater social, economic and cultural equality, there is a vital role for alternative or counter-public spheres where subordinate or marginalized groups can caucus on their own, as it were, in order to identify their own interests, build a culture, marshal resources and formulate strategies for change. Then, they can re-engage with the broader, general arenas of the media and political systems where 'public opinion' is in theory articulated (Fraser 1997).

Saving the baby

Such a combative notion of public spheres sharply qualifies more liberal versions. Yet we need not discard the conceptual baby with the public sphere's historical and political bathwater. The public sphere's historical exclusions (of women, workers, non-citizens) and the politically naïve expectations of some of its advocates do not eliminate its utility as an ethical horizon – one never arrives there, but it provides a direction in which to walk (Dahlgren 1995: 118). Three aspects are worth maintaining in relation to the imperative of mobilizing people for just and effective responses to climate crisis.

First, the public sphere is in principle open to everyone. It has an immanent logic of inclusion appropriate for ensuring that creative as well as obstructionist voices, the 'voices of the side effects' as well as the biggest greenhouse gas emitters, are heard.

Second, the public sphere is conceptually linked to the use of reason and evidence to adjudicate disagreements. Reason and evidence would be welcome counterweights to the deliberate misinformation generated by climate science denialists (Oreskes and Conway 2010) and the 'willful blindness' of governments such as Stephen Harper's in Canada and Tony Abbott's in Australia – governments that muzzled scientists and defunded research on politically inconvenient topics such as global warming (Turner 2013).

Third, an effective public sphere is guided by a search for a public interest broader than private interests. The massive public relations efforts by fossil fuel companies such as Enbridge, seeking public support for its proposed Northern Gateway pipeline from the Alberta tar sands to the Pacific coast, should dispel the naïve assumption that private interests can easily be bracketed out from public deliberation. Instead, private interests should be declared and made transparent, and the economic and cultural resources for engagement in public decision-making should be evened out between the parties involved. That levelling of resources implies political and social changes to maintain an 'enabling environment' for meaningful deliberation, one that carries beyond decision-making processes themselves.

Taken together, the characteristics of inclusiveness, reason, and transparency, when bolstered by social equality, should help offset one of the main problems of many actually existing public spheres (such as the mass media) – their tendency to privilege the interests and perspectives of those who already dominate the social and political systems. Through the ideological work of political parties, policy institutes, state agencies and mass media in particular, those groups have disproportionate power to define risks, crises, common sense and the public interest. They can be considered the 'primary definers' of social reality (Hall *et al.* 1978). Nancy Fraser's model of counter-public spheres, where excluded groups can forge alliances and define their interests before re-engaging with the broader public sphere, suggests one kind of strategy to counteract such hegemonic power. The alternative media discussed in this book can be seen as helping to constitute such counter-spheres.

One of the advantages of the public sphere as a normative model is its convertibility to different scales. Historically, it has been conceptualized as operating at the level of national states, which are still key sites where struggles for effective climate and sustainable energy policies are being fought. The globalization of the social order raises the question of how public spheres can be transposed to transnational arenas where cultural, ethnic, national and class contradictions co-exist with the homogenizing and ecologically destructive logic of capital accumulation, but where there is no single centralized authority despite the urgent necessity of coordinated action.

We do not yet know what genuinely global public spheres would look like. Transnational social movements with varying degrees of spontaneity and cohesiveness – Social Forum, Occupy, Arab Spring – are one possible model. Nongovernmental organizations such as Amnesty International, with its active chapters and membership of several million, are a second possibility; arguably, they have nudged global politics in the direction of 'monitory democracy' (Keane 2009), something that falls far short of participatory democracy but registers the presence of emerging global public opinion in the calculations of governments. It may be that a global public sphere would resemble a network of smaller-scale public spheres, which may not necessarily need to be formally linked in institutional decision-making bodies. It is at the local community level where public spheres can most easily take shape, and carry out campaigns and actions that are exemplary and inspirational, and that resonate elsewhere, potentially globally (Foxwell-Norton 2013). One example: The environmental movement in British Columbia received a substantial boost from a referendum in Kitimat, a small coastal town where the Enbridge Corporation had heavily touted the alleged economic benefits of its proposed pipeline from the Alberta tar sands. After fractious debates, residents nevertheless voted by 60 to 40 percent to reject it.

Fractious indeed. Any public sphere that would be adequate to the task of addressing global ecological crisis is an unavoidably contentious space. Even in a political unit as cohesive and modest in scale as British Columbia, a forum that gives voice to constituencies as diverse as carbon industry capitalists, construction unions, Indigenous peoples, environmental groups, recreational land and water users, city governments, household owners and others is likely to be something other than a 'genteel conversation' over issues such as pipeline construction. Scale that up to the global level, and the challenge is multiplied. So public spheres in global context are likely to be 'a series of embattled fields of contention, insurgency and reflexivity that are local to transnational in scope' (Cottle 2009: 38, citing Dryzek 2006). Consensus may not be achievable or even desirable, but at their best, representative and effective public spheres offer hope that arguments carry greater weight than armaments.

The contentiousness of public spheres in relation to climate crisis shifts us from liberal/deliberative to radical conceptions of democracy. Interestingly, Amanda Machin (2013) asks how 'radical democracy' might apply to negotiating climate change. Machin critiques what she sees as the dominant approaches to climate politics. The 'techno-economic' model places too much faith in science and markets as magic bullets. The 'ethical-individual' approach unrealistically expects ethical consensus, and deflects responsibility from governments and corporations (p. 42). These two approaches are similar to the neoliberal response to climate change already sketched. But Machin also attacks green variants of public sphere liberalism. 'Green republicanism' is an advance over individualist approaches and acknowledges the value of community, active citizenship and the 'common good' (p. 47); but ultimately, these categories imply a universalizing override of diversity, tending towards authoritarianism. 'Green deliberative democracy' is more promising because it celebrates a 'myriad of differences in a society' (p. 68), and what is needed is more democracy, not less, because:

> environmental concerns are excluded from a system in which a privileged elite are able to insulate themselves from the impact of environmental destruction, have the greatest interest in the continuation of the status quo, and are the very people in control of decision-making.
>
> (p. 73)

Yet (Machin argues), deliberative democracy falls short because it tries to impose general norms or processes of deliberation that marginalize some people's speech (pp. 80–81).

Machin's analysis has its own limitations. Centrally, she rejects consensus as an undesirable and unattainable goal (her book's subtitle is 'the illusion of consensus'), yet repeatedly rejects other approaches as insufficiently inclusive. She acknowledges the influence of power imbalances on public deliberation, but is oddly dismissive of the organized social movements that could potentially redress them – on the grounds that movements engage in instrumentalist communication towards predetermined goals (p. 104). An excessive concern with the inclusion/exclusion of voices begs the question of whether all 'differences' actually deserve to be respected and included. Does the Ku Klux Klan deserve a voice on race relations, or are they beyond the 'frontier'? Moreover, a continual process to ensure inclusion of all possible voices invites a continual deferral of decisive action (Schultz 2014: 369). Such deferral not only means victory by default for a high-carbon status quo and the powerful interests that perpetuate it but also it escalates the risks of crossing various biophysical tipping points, and losing a race against time.

Nevertheless, radical democratic theory has important implications for climate politics strategies – the celebration rather than the avoidance of conflict, the need to counterbalance power relations that can influence policy, and the potentially invigorating role of open disagreement (a theme explored in detail in Chapter Five). These are advances over consensus-oriented public sphere liberalism, especially in an area as contentious as climate politics.

Radical conceptions of democracy and the public sphere appear to imply a hybrid radical facilitative role for journalism. In part, journalism would facilitate communicative space for excluded and/or oppositional voices, as Fraser's model of subaltern public spheres suggests, and it would actively engage in developing critiques of established power. What would be the content of those critiques? What substantive platforms, ideologies or philosophies might inform pro-climate politics and related strands of journalism? The answer is very much contingent on the ebb and flow of struggles, the creativity and agency of transformative social movements, conversations that occur in threatened communities, the work of public intellectuals, and much else.

Oppositional themes

This section is intended only to illustrate three of the themes that have emerged in recent years as people strive to define and realize a more sustainable and hopeful future. These themes are generally ignored or demonized in corporate media, but they could provide frames for both journalism and public discussion in open, diverse, democratic and globally-oriented public spheres.

The rights of nature

One positive approach to ecological politics was prefigured by the small South American country of Ecuador, which in 2008 became the first nation constitutionally to enshrine the rights of nature 'to exist, persist, maintain and regenerate its vital cycles, structure, functions and its processes in evolution', defining nature as 'a rights-bearing entity that should be treated with parity under the law' (Phillips and Huff 2009: 83). The 'rights of nature' could be pushed to logical extremes: Should environmentally harmful industry and agriculture be abruptly ended, or endangered species protected, even if it costs millions of human lives? A more politically viable vision of the rights of nature sees it not as a means to prioritize non-human species, but as a means to sustain humanity's own habitat, to let long-term stewardship trump short-term economic exploitation, and to reinforce cultural narratives more sustainable than the Enlightenment and capitalist myth of eternal expansion and domination over nature.

Obviously, nature's legal and political voice requires human proxies, such as Indigenous peoples with intense relationships to the land, and this necessitates more democracy, not less. Parallel to that constitutional commitment, the centre-left government of Rafael Correa in 2007 offered to protect the incredibly bio-diverse Yasuni National Park from extractivist exploitation if the international community would provide half the estimated value of the underlying oil deposits. Unfortunately, much less than one percent of the $3.6 billion goal, intended to help Ecuador pull itself out of poverty, was raised. In 2013, the Correa government announced it would permit drilling, sparking an internal opposition movement to protect the forest (Klein 2014: 410–1).

The Ecuadorean struggle suggests the need for both material resources and strong social movements to combat the logic of extractivist capitalism. Constitutional phrases are not enough. If avoiding climate catastrophe requires that countries in the global South refrain from fully exploiting their fossil fuel deposits or following the same carbon-intensive development that has fuelled the global North's economic prosperity, then the South needs to be offered an alternative route to escape poverty. Ecuador's constitutional rights of nature (echoed by Bolivia and Mongolia, by the 2010 World Peoples' Conference on Climate Change and the Rights of Mother Earth, and by subnational jurisdictions in Europe and North America [Klein 2014: 444]) imply a new task, and perhaps an alternative rationale, for democracy – obtaining consent and legitimacy for the shared sacrifices that societies, particularly in the global North, need to make in the interests of a sustainable common future.

Climate justice

One potentially powerful tool in that task is 'climate justice'. It is both a concept and a growing transnational movement, based on the premise that those who have most benefitted from and/or contributed to atmospheric dumping should shoulder most of the burden of redressing the crisis – especially since those most vulnerable to the negative impacts of climate change (drought, famine, extreme weather) had the least responsibility for creating them. The academic literature on climate justice manifests a variety of approaches and emphases. For example, what is it intended to achieve? How does one separate the responsibilities and obligations of individuals from those of whole countries and of current generations from past ones? How does climate justice relate to other concepts of justice – distributive, procedural, corrective, intergenerational, global? Should it be integrated with them, or developed separately? How important is it that the normative ideal of climate justice be readily translatable into substantive policy within existing institutions? We shall bypass these debates (but for some overviews and interventions, see Bell 2013; Cox 2010: 281–7; Goodman 2010; Posner and Sunstein 2008). Climate justice is discussed further in Chapter Two; here, I highlight aspects that might be most relevant to journalistic frames and practices.

In a special issue in January/February 2009, *New Internationalist* magazine outlined four principles for the practical implementation of climate justice. First, the rich – the already industrialized countries and wealthy elites within both global North and South – must take primary responsibility for the burdens of adjustment, given their role in creating and perpetuating the crisis. This means drastically cutting their greenhouse gas emissions, ending overproduction for overconsumption, and financially helping the global South to adjust to climate change impacts and to develop along more sustainable lines. Second, most fossil fuels need to be left in the ground, and investment re-directed to energy efficiency and community-controlled renewable energy. Third, solutions need to be both fair and effective, protecting people's rights, jobs and well-being, and avoiding 'false solutions' like carbon trading, biofuels and geo-engineering. Finally, natural resources should be conserved for the common good, not private or unsustainable exploitation, and people's sovereignty over them reclaimed (cited by Lee 2009).

Justice may seem to be outside the realm of 'realpolitik'; it has not typically impacted intergovernmental climate policy discussions dominated by horse-trading negotiations (Bell 2013). Yet climate change may be a case where Might needs Right to succeed. How else could newly industrializing countries like India, already victimized by melting glaciers, droughts and unprecedented heat waves, be persuaded to abandon the same developmental path as the high-emission global

North? Unlike most other issues, even the powerful need a genuinely global solution that cannot be achieved without an engagement with justice (Baskin 2009).

There is no denying, though, that climate justice would be a difficult sell amongst the electorates of the global North, if it calls for reduced energy consumption and increased taxation. It might need not just a fundamental political shift, but also a cultural one. A culture that respected both justice and nature would need a subjectivity different from that cultivated by neoliberalism and consumerism – the self-seeking, utility-maximizing autonomous individual that is assumed and celebrated by neoclassical economics, the ideological underpinning of neoliberalism. Through regimes of economic production, as well as media and other institutions of socialization (schools, religion, the family), social systems tend to produce particular personality types – a 'character structure which is shared by most members of the same culture' that serves 'to mold and channel human energy within a given society for the purpose of the continued functioning of this society', as social philosopher Erich Fromm put it (1955: 76–7). It may be that neoliberal versions of capitalism are more likely than other systems to produce the personality type colloquially known as the Asshole – individuals (overwhelmingly male) with an excessive sense of entitlement that renders them both willing to exploit or bully others, and immune to criticism or correction (James 2012: 4–5).

Journalism, and media in general, are not neutral and innocent, no matter how 'objective' they claim to be. American media scholar George Gerbner pioneered research on how dominant mass media comprise a symbolic environment that 'cultivates' perceptions of reality and attendant attitudes; heavy viewers of violent television drama were more likely to see the world as threatening, and to favour authoritarian approaches to social problems (Gerbner and Morgan 2002). The French Marxist philosopher Louis Althusser (1971: 127–86) wrote of how 'ideological state apparatuses' like media 'interpellate' or address people, inviting them to define themselves and situate themselves in the world in particular ways – to adopt particular subject positions. Such analyses have been criticized as too one-sided and functionalist, underestimating audience resistance, and they have been supplanted by recent theorizing on media and subjectivity (Corner 2011). But they do suggest one plausible ethos for journalism in the context of climate crisis: Don't normalize or encourage Assholery. Instead, media could become part of what James (2012) labels the culture's 'Asshole Dampening System', the institutions such as family and education that societies need to maintain civic trust and cohesion. Media could present active citizenship, community engagement and shared sacrifice as normal. They could offer more stories of people working together to make positive political changes and fewer SUV ads or Donald Trump-style 'reality' television shows.

Ecosocialism

Assholes are not produced at random. Personality types, the distortion of public deliberation by power imbalances and the unfolding climate crisis are all arguably products of a system. The socialist tradition (whose revitalization is evident in the emergence of politicians like former U.S. presidential candidate Bernie Sanders and British Labour Party leader Jeremy Corbyn) has named that system – capitalism – and held it up for analysis and transformation in relation to global climate emergency. Ecosocialists argue that capitalism's inherent drive for endless capital accumulation is the key force behind the ongoing destruction of humanity's planetary habitat. Consequently, in this view, capitalism and civilization survival are ultimately incompatible (see e.g. Kovel 2007; Magdoff and Foster 2011). Moreover, the drive to exploit labour and nature links ecological destruction with poverty and inequality, and the brunt of climate change falls disproportionately on the masses of the global South. The ecosocialist slogan 'system change, not climate change' might have seemed wacko a few years ago, even within the progressive environmental movement, but it now strikes much louder chords. One prominent example is Pope Francis's 2015 encyclical *Laudato Si: On Care for Our Common Home*. While the Pope is no Marxist, and his remarkable document does not specifically mention 'capitalism', it denounces 'the machinery of the current globalized economy' as 'a system of commercial relations and ownership which is structurally perverse' in its impact on both natural environment and human dignity. The encyclical identifies 'obsession with unlimited growth, consumerism, technology, the total domination of finance and the deification of the market' as systemic problems and links poverty and environmental degradation (Löwy 2015: 50–1).

Rethinking journalism ethics

Given that journalism is inherently a political practice, what do the political vectors just discussed entail for journalism's democratic ethos? The argument is not that journalism should become an exclusionary or propagandistic outlet for particular ideas or movements, but rather that journalism needs to be far more open to the kinds of ideas just discussed than conventional media currently are.

In addition to its crisis of political economy, Western dominance of journalism ethics has been under challenge from both outside and within the media field. Postmodernism has expressed deep scepticism about the possibility of truth and objectivity; postcolonial critics see the hidden fist of Western power embedded in apparently universal values; and a new technological and social environment has forged a 'new media' journalism with different values, such as openness and transparency rather than professionally defined objectivity and social responsibility (Ward 2009: 301–4). In the wake of the anti-colonial revolutions of the decades following World War II, new conceptions of communication rights addressed such 'planetary concerns' as peace, development, communication and ecological balance. Such rights can only be implemented through international cooperation; they entail a duty for states and social organizations 'to place common human interests before national and individual interest' (Traber 1993: 24–5).

The exigencies of climate crisis resonate with some of these new turns in journalism ethics – global vision, cultural sensitivity, collective and interdependent interests. I conclude this chapter by reconsidering the normative roles of journalism in light of the foregoing analysis.

The monitorial role

The *monitorial* role – the reporting of issues; the role of watchdog on power; informing media audiences of policies, actions and events that affect them; scanning the scene for threats to individual, community and societal well-being – is acknowledged in all the previously mentioned models of democracy. It becomes all the more relevant in light of climate crisis.

Indeed, 'crisis' could be a defining feature of climate politics journalism. Scholar Robert Cox (2007: 15–16) argues for making environmental communication a 'crisis discipline', one that is inherently political and public-oriented. He offers the beginnings of a model, based on four ethical principles:

1   'Enhance the ability of society to respond appropriately to environmental signals' for the benefit of human and environmental health
2   Make relevant information and decision-making processes 'transparent and accessible to members of the public' while those affected by environmental threats 'should also have the resources and ability to participate in decisions affecting their individual or communities' health'
3   Engage various groups 'to study, interact with and share experiences of the natural world'
4   Critically evaluate and expose communication practices that are 'constrained or suborned for harmful or unsustainable policies toward human communities and the natural world' (Cox 2007: 15–16).

The first two at least, fit well with journalism's democratic role of monitoring power and surveying the social and physical environment for well-being. The second and fourth entail a critical distance from established power, and a sensitivity to the rights and voices of the 'side effects' of ecological crisis (Beck 1992). Those themes parallel the ethos of climate justice, which could well provide a meta-frame for climate crisis reporting. Climate justice suggests a reflexive and critical monitoring of climate policies and impacts. It invokes greater attention to climate disruption in the most vulnerable parts of the world, to the question of who benefits or suffers from high-emission economic development, and to who should pay the costs of transition to a low-carbon economy. It inspires linkage between events that conventional journalism typically compartmentalizes, such as free trade agreements and locally sustainable economies or climate activism with other social justice movements. And it posits climate crisis as an ethical question, not just another political controversy.

But there is some debate about whether monitorial reporting that is inspired by a meta-frame, and a politically loaded one at that, could – or should – be objective, and what objectivity means in practice. At first, it seems obvious that news should be presented without contamination by personal opinion or vested interests. Nowhere, it could be argued, is that more important than in dealing with global crises like climate change. If the truth simply be told, surely we would act. Yet, beyond the assumption (critiqued in Chapter Two) that knowledge necessarily leads to action, objectivity – as it has been defined in dominant news media – has been subjected to withering critiques (Maras 2013). The least important of these is its alleged impossibility if objectivity means the exclusion of journalists' individual 'biases' that may lead them to 'distort' accurate representations of the real. More telling are critiques that see objectivity as an insufficient ethical guide – as Chris Wood, one of our inter-view respondents in Vancouver put it (see Chapter Five), objectivity indicates how but not what 'stories' should be 'covered' – or as deceptive, presenting as neutral and universally valid accounts that upon close analysis prove to be ideologically loaded, usually favouring dominant interests, power relations or worldviews.

But others argue that the stream of critique has gone on for too long, that the task is no longer to deconstruct journalism but to reconstruct it (e.g. Calcutt and Hammond 2011). Arguably, journalism is in need of rescuing from a morass of self-serving public relations, postmodernist relativism and hyperventilating pun-dits whose pontifications are seldom scrutinized against political reality. Real-ity checks become all the more important given the prominence in U.S. media accorded to climate change sceptics who are often linked to industry-funded cam-paigns to muddy public opinion and delay effective policy responses (Boykoff and Boykoff 2004; Oreskes and Conway 2010). Communication that informs and mobilizes publics to engage effectively with climate change needs to address its biophysical realities, to distill and convey what scientists have been warning humankind about. In that sense, objective reporting serves a coincidence of pur-poses between journalism, science and survival.

We conclude that certain practices that have been considered characteristic of 'objective' journalism and that have been highly problematic for media represen-tations of environmental crisis need to be abandoned or replaced (artificial 'bal-ance', overemphasis on official/political sources, privileging of observable events over long-term processes and structures) in the interest not of abandoning the tell-ing of publicly relevant truths but rather telling them more accurately, completely, pluralistically and inclusively.

The facilitative role

From the perspective of a robust, crisis-oriented public sphere, key tasks for journalism include stimulating interest in and attention to global risks, clarify-ing arguments, rendering hidden interests transparent, providing fair access to the contending perspectives regardless of their economic resources, and *facilitat-ing* public conversation, a search for solutions and workable agreement – if not consensus – on remedial action.

In those tasks, journalism does not need to reinvent the wheel. Theoretical foundations for this view of journalism and its publics were laid a century ago by the American pragmatist philosopher John Dewey (1927). Consider Stuart Allan's (2010) interpretation of Dewey's argument:

Dewey is discerning in the press a capacity for social reform. . . . The journalist, like the social scientist, is charged with the responsibility of providing the information about pressing issues of the day – as well as interpretations of its significance – so as to enable members of the public to arrive at sound judgements. . . . (I)t seems apparent to Dewey that democracy must become more democratic, that is, more firmly rooted in everyday communities of interaction. To the extent that the journalist contributes to the organization of the public – not least by facilitating lay participation in the rough and tumble of decision-making – the citizenry will be equipped to recognize, even challenge the authority exercised by powerful interests.

(p. 68)

Although Dewey wrote four decades before 'the environment' emerged as a field of media representation and political contestation, his analysis prefigures the tasks that many environmentalists see as critical for contemporary journalism: move beyond simply informing publics about climate crisis in order to motivate popular engagement and help build the political capacity to address crisis. That capacity is multifaceted: it includes the formation of publics out of the atomization and isolation that people experience in consumer capitalism; their willingness and ability to participate (through both expressing themselves and taking into account the perspective of others) in the discussion of public issues; the ability to evaluate information and assess the relative risk of different policy options; trust in the potential of collective remedial action; a growing sense of political efficacy; and mechanisms by which popular opinion can not only challenge powerful interests, including the fossil fuel industry, but also set the agenda of policy-making. Mass-based political capacity is a resource that can help civil society address not only climate change mitigation and adaptation, but also other environmental and social crises and challenges. Not only social movements, political parties and advocacy groups, but also journalism, have distinct potential roles in such capacity building.

How could journalism contribute to public engagement and capacity building? Much remains to be learned about these processes, but there are practical forebears. The alternative media discussed in subsequent chapters offer examples of journalism that both challenges dominant frames and builds counter-public spheres; many of their practitioners see themselves as stimulating engagement and forging options for effective community responses to such challenges as climate change. Even corporate media have also sometimes experimented with promoting public engagement in relation to conflicts and community issues. Two well-organized and theorized reform efforts – Peace Journalism and Civic Journalism – are discussed in Chapter Four. They may offer particular lessons for climate crisis reporting.

The radical role

While still relatively marginal in the journalism field, the *radical* role also has its place in actively advocating and mobilizing for fundamental change against the resistance of hegemonic institutions. If there is any validity to the arguments previously noted, that the fossil fuel industries are structurally an enemy of effective climate policy and/or that capitalism has environmentally destructive expansion built into its very dynamics, then journalism for climate crisis also requires structural independence from those forces and a receptiveness to oppositional movements for change. A press system that is dependent on corporate advertising, or on attracting online clicks for constructing commercially valuable consumer profiles is not likely to offer that kind of discursive space.

That does not imply that journalism should become an adjunct of political movements, in the manner of the Leninist model of the press as an instrument of a vanguard party, leading the charge towards the Promised Land. Under conditions of modernity and democracy, journalism has a societal function, a capacity for an independent gaze that separates it from propaganda (Calcutt and Hammond 2011). But in relation to climate politics, the radical role could entail advocacy journalism, the provision of 'mobilizing information' to encourage popular participation in anti-pipeline campaigns, and the framing of information through a counter-hegemonic lens such as climate justice.

In terms of the political positions previously discussed, these kinds of journalism roles are obviously more aligned with radical democracy and ecosocialism than with market liberalism or the spuriously neutral stance of deliberative democracy. Structurally, they are more likely to be nurtured within the independent and alternative press rather than corporate- or state-owned media.

The collaborative role

Nevertheless, the *collaborative* role, while similarly marginalized in the liberal tradition, might also have a place in supporting other institutions' productive environmental campaigns and policies – from community recycling to a nationwide carbon tax. Conventional critics might scoff at journalistic boosterism, but it is difficult to imagine successful climate mitigation and green energy policies without vigorous support in the public arena. And it is hardly alien in the practice of Western journalism, for better or worse. Weekly community newspapers are the most obvious venue for journalism that actively supports local initiatives – charity campaigns, festivals and the like. But mainstream journalism, without admitting it, often resembles collaboration, from hometown sports teams, to support for 'our troops' in wartime – and even for the governmental policies that led to war, most notoriously, U.S. media acceptance of the rationales that George W. Bush offered for invading Iraq in 2003.

If our overriding concern is with turning human society sharply away from falling off the climate cliff, as distinct from pursuing democratic principles abstracted from specific political issues, then what models or roles are to be preferred? Research in environmental communication helps to identify what kinds of communication strategies and frames are most likely to effectively mobilize public engagement. Attention turns to that question in Chapter Two.

Note

1 Portions of this section derive from Hackett (2005), Hackett and Carroll (2006: 69–74), and Zhao and Hackett (2005).

References

Allan, S. (2010) 'Journalism and its publics: The Lippmann-Dewey debate', in S. Allan (ed), *The Routledge Companion to News and Journalism*, London and New York: Routledge, pp. 60–70.

Althusser, L. (1971) *Lenin and Philosophy and Other Essays*, New York: Monthly Review Press.

Arias-Maldonado, M. (2007) 'An imaginary solution? The green defence of deliberative democracy', *Environmental Values* 16(2): 233–52.

Baker, C.E. (2002) *Media, Markets and Democracy*, Cambridge, UK: Cambridge University Press.

Baskin, J. (2009) 'The impossible necessity of climate justice?', *Melbourne Journal of International Law* 10(1): 424–38.

Beck, U. (1992) *Risk Society: Towards a New Modernity*, Newbury Park, CA: Sage.

Beeson, M. (2010) 'The coming of environmental authoritarianism', *Environmental Politics* 19(2) (March): 276–94.

Bell, D. (2013) 'How should we think about climate justice?', *Environmental Ethics* 35(2): 189–208.

Boykoff, M.T. and Boykoff, J.M. (2004) 'Balance as bias: Global warming and the US prestige press', *Global Environmental Change* 14(2): 125–36.

Calcutt, A. and Hammond, P. (2011) *Journalism Studies: A Critical Introduction*, Abingdon, UK, and New York: Routledge.

Carpentier, N. and Cammaerts, B. (2006) 'Hegemony, democracy, agonism and journalism: An interview with Chantal Mouffe', *Journalism Studies* 7(6): 964–75.

Carter, N. (2007) *The Politics of the Environment: Ideas, Activism, Policy*, 2nd edn, Cambridge, UK: Cambridge University Press.

Christians, C., Glasser, T., McQuail, D., Nordenstreng, K. and White, R. (2009) *Normative Theories of the Media: Journalism in Democratic Societies*, Urbana and Chicago: University of Illinois Press.

Corner, J. (2011) *Theorising Media: Power, Form and Subjectivity*, Manchester, UK: Manchester University Press.

Cottle, S. (2009) *Global Crisis Reporting: Journalism in the Global Age*, Maidenhead, UK: Open University Press.

Cox, R.J. (2007) 'Nature's "crisis disciplines": Does environmental communication have an ethical duty?', *Environmental Communication: A Journal of Nature and Culture* 1(1): 5–20.

———— (2010) *Environmental Communication and the Public Sphere*, 2nd edn, Los Angeles: Sage.

Cunningham, F., Smelser, N. and Baltes, P. (2015) 'Democratic theory', in *International Encyclopedia of the Social and Behavioral Sciences*, 2nd edn, Amsterdam: Elsevier, pp. 90–6.

Curran, J. (2011) *Media and Democracy*, London and New York: Routledge.

Dahlgren, P. (1995) *Television and the Public Sphere: Citizenship, Democracy and the Media*, London: Sage.

Dewey, J. (1927) *The Public and Its Problems*, Athens, OH: Swallow Press.

Dolack, P. (2015) 'Business as usual at Paris summit won't stop global warming'. Accessed at http://www.counterpunch.org/2015/12/18/business-as-usual-at-paris-summit-wont-stop-global-warming/.

Downing, J.D.H., with Ford, T.V., Gil, G. and Stein, L. (2001) *Radical Media: Rebellious Communication and Social Movements*, Thousand Oaks, CA: Sage.

Dryzek, J. (2006) *Deliberative Global Politics*, Cambridge, UK: Polity.

Forde, S. (2011) *Challenging the News: The Journalism of Alternative and Community Media*, Houndmills, UK, and New York: Palgrave Macmillan.

Foxwell-Norton, K. (2013) 'Communication, culture, community and country: The lost seas of environmental policy', *Continuum: Journal of Media and Cultural Studies* 27(2): 267–82.

Fraser, N. (1997) *Justice Interruptus*, New York: Routledge.

Fromm, E. (1955) *The Sane Society*, New York: Fawcett.

Gerbner, G. and Morgan, M. (2002) *Against the Mainstream: The Selected Works of George Gerbner*, New York: Peter Lang.

Giddens, A. (2009) *The Politics of Climate Change*, Cambridge, UK: Polity.

Goodman, J. (2010) 'From global justice to climate justice? Justice ecologism in an era of global warming', *New Political Science* 31(4): 499–514.

Gore, A. (2006) *An Inconvenient Truth*, Rodale, PA: Emmaus.

Gramsci, A. (1971) *Selections from the Prison Notebooks*, New York: International Publishers.

Hackett, R.A. (2005) 'Is there a democratic deficit in US and UK journalism?', in S. Allan (ed), *Journalism: Critical Issues*, New York: Open University Press/McGraw-Hill, pp. 85–97.

Hackett, R.A. and Carroll, W.K. (2006) *Remaking Media: The Struggle to Democratize Public Communication*, New York and London: Routledge.

Hackett, R.A. and Gurleyen, P. (2015) 'Beyond the binaries? Alternative media and objective journalism', in C. Atton (ed), *The Routledge Companion to Alternative and Community Media*, London: Routledge, pp. 54–65.

Hall, S., Critcher, C., Jefferson, T., Clarke, J. and Roberts, B. (1978) *Policing the Crisis: Mugging, the State, and Law and Order*, London: MacMillan.

Hallin, D.C. and Mancini, P. (2004) *Comparing Media Systems: Three Models of Media and Politics*, Cambridge, UK: Cambridge University Press.

Hansen, A.D. and Sonnichsen, A. (2014) 'Radical democracy, agonism and the limits of pluralism: An interview with Chantal Mouffe', *Distinktion: Scandinavian Journal of Social Theory* 15(3): 263–70. DOI: 10.1080/1600910X.2014.941888.

Held, D. (2006) *Models of Democracy*, 3rd edn, Stanford, CT: Stanford University Press.

hooks, b. (1981) *Ain't I a Woman? Black Women and Feminism*, Boston: South End Press.

James, A. (2012) *Assholes: A Theory*, New York: Doubleday.

Josephi, B. (2013) 'Decoupling journalism and democracy: How much democracy does journalism need?', *Journalism* 14(4): 441–5.

Keane, J. (2009) *The Life and Death of Democracy*, London and New York: Simon and Schuster.

Kellner, D. (2003) *From 9/11 to Terror War: The Dangers of the Bush Legacy*, Lanham, MD: Rowman & Littlefield.

Kenis, A. and Lievens, M. (2014) 'Searching for "the political" in environmental politics', *Environmental Politics* 23(4): 531–48.

Klandermans, B. (2001) 'Why social movements come into being and why people join them', in J. Blau (ed), *Blackwell Companion to Sociology*, Malden, MA: Blackwell, pp. 268–81.

Klein, N. (2014) *This Changes Everything: Capitalism vs the Climate*, Toronto: Knopf.

Kovel, J. (2007) *The Enemy of Nature: The End of Capitalism or the End of the World?*, New York: Zed Books.

Laclau, E. and Mouffe, C. (1985) *Hegemony and Socialist Strategy: Towards a Radical Democratic Politics*, London: Verso.

Lee, P. (2009) *The No-Nonsense Guide to Communication, Climate Justice and Climate Change*, Toronto: World Association for Christian Communication. Accessed at http://www.wacglobal.org/en/activities/climate-justice.html.

Lippmann, W. (1922) *Public Opinion*, New York: Free Press.

Lovelock, J. (2010, March 30) Interviewed by L. Hickman of *The Guardian*. Accessed at http://www.theguardian.com/environment/blog/2010/mar/29/james-lovelock.

Löwy, M. (2015) 'Laudato Si – the pope's anti-systemic encyclical', *Monthly Review* 67(7) (December): 50–4.

Machin, A. (2013) *Negotiating Climate Change: Radical Democracy and the Illusion of Consensus*, London: Zed Books.

Macpherson, C.B. (1966) *The Real World of Democracy*, Oxford, UK: Oxford University Press.

—— (1977) *The Life and Times of Liberal Democracy*, Oxford, UK: Oxford University Press.

Magdoff, F. and Foster, J.B. (2011) *What Every Environmentalist Needs to Know about Capitalism*, New York: Monthly Review Press.

Maras, S. (2013) *Objectivity in Journalism*, Cambridge, UK: Polity Press.

Martell, L. (1994) *Ecology and Society*, Cambridge, UK: Polity Press.

McKibben, B. (2012) 'Global warming's terrifying new math', *Rolling Stone*. Accessed at http://www.rollingstone.com/politics/news/global-warmings-terrifying-new-math-20120719.

Miliband, R. (1973[1969]) *The State in Capitalist Society: An Analysis of the Western System of Power*, London: Quartet Books.

Mouffe, C. (2002) 'Which public sphere for a democratic society?', *Theoria: A Journal of Social and Political Theory* 99 (June): 55–65.

Niemeyer, S. (2013) 'Democracy and climate change: What can deliberative democracy contribute?', *Australian Journal of Politics and History* 59(3): 429–48.

Norris, P. (2000) *A Virtuous Circle: Political Communications in Postindustrial Societies*, Cambridge, UK: Cambridge University Press.

Oreskes, N. and Conway, E. (2010) *Merchants of Doubt: How a Handful of Scientists Obscured the Truth on Issues from Tobacco Smoke to Global Warming*, New York: Bloomsbury Press.

Pew Research Center (2012) *Global Attitudes Project*. Accessed at http://www.pewglobal.org/2012/05/23/chapter-3-attitudes-toward-democracy-2/.

Phillips, P. and Huff, M. with Project Censored (eds) (2009) *Censored 2010: The Top 25 Censored Stories*, New York: Seven Stories Press.

Posner, E.A. and Sunstein, C.R. (2008) 'Climate change justice', *Georgetown Law Journal* 96(5): 1565–612.

Schultz, P. (2014) Review of *Negotiating Climate Change: Radical Democracy and the Illusion of Consensus* by A. Machin, *Environmental Philosophy* 11(2): 366–69.

Schumpeter, J. (1976) *Capitalism, Socialism and Democracy*, London: Allen and Unwin.

Shearman, D.J.C. and Smith, J.W. (2007) *The Climate Change Challenge and the Failure of Democracy*, Westport, CT: Praeger.

Traber, M. (1993) 'Changes of communication needs and rights in social revolutions', in S. Splichal and J. Wasko (eds), *Communication and Democracy*, Norwood, NJ: Ablex, pp. 19–31.

Turner, C. (2013) *The War on Science: Muzzled Scientists and Wilful Blindness in Stephen Harper's Canada*, Vancouver: Greystone Books.

Ward, S.J.A. (2009) 'Journalism ethics', in K. Wahl-Jorgensen and T. Hanitzsch (eds), *Handbook of Journalism Studies*, New York: Routledge, pp. 295–309.

Young, I.M. (2001) 'Activist challenges to deliberative democracy', *Political Theory* 29(5): 670–90.

Zhao, Y. and Hackett, R.A. (2005) 'Media globalization, media democratization: Challenges, issues, and paradoxes', in R.A. Hackett and Y. Zhao (eds), *Democratizing Global Media*, New York: Rowman & Littlefield, pp. 1–33.

Chapter 2

Engaging climate communication

Audiences, frames, values and norms[1]

Shane Gunster

Engagement, efficacy and the limits of information

Most communication about climate change generally assumes that providing accurate information about its causes, consequences and solutions will necessarily improve public awareness and understanding of the problem, generate increased levels of care and concern, and motivate greater support for policy solutions and behavioural change. From *An Inconvenient Truth* to media coverage of scientific reports and policy summits to advocacy campaigns, much climate change communication has been organized around the simple belief that 'members of the lay public need to be better informed about climate change in order for them to engage in pro-environmental practices. . . . [P]eople are rational, responsible actors who simply require appropriate information in order to alter their behaviours and support policy change' (Potter and Oster 2008: 119). Once people learn about the magnitude, severity and urgency of the problem, levels of concern and willingness to take action will necessarily rise. Likewise, public indifference or reluctance to support aggressive policies is often attributed to a 'deficit' of relevant information, and the failure (or refusal) of governments, journalists and others to give this issue the attention it deserves. Perception of the need to mitigate this deficit has likely been intensified by the fact that despite broad public awareness of climate change, overall levels of knowledge and understanding remain quite poor (Nisbet and Myers 2007). Summarizing two decades of U.S. opinion polls, Matthew Nisbet and Teresa Myers note that 'few Americans are confident that they fully grasp the complexities of the issue, and on questions measuring actual knowledge about either the science or the policy involved, the public scores very low' (2007: 447; also see Leiserowitz *et al.* 2010). For those seeking greater public engagement with climate change, addressing gaps and deficiencies in knowledge seems like an obvious priority.

Susanne Moser, a leading scholar and practitioner in environmental communication, attributes this deficit of knowledge and engagement to key features of climate change (and our interactions with it) which make it particularly challenging to understand and care about, including: invisible causes (connections between activities, emissions and impacts are highly abstract and often imperceptible in everyday life); geographically and temporally distant impacts; insulation of modern humans from their environment ('direct' experiences of climate impacts are rare and often not perceived as such); delayed or absent gratification for taking action (substantial lag between mitigation actions and beneficial effects); homo sapiens' brain versus homo technologicus's power (difficulty in recognizing that individual actions have an aggregate impact upon global climatic systems); complexity and uncertainty (the specialized discourses of climate science and policy are difficult for many to understand and often subject to wilful misinterpretation); inadequate signals indicating the need for change (e.g. weak or non-existent carbon pricing insulates consumers and businesses from market-based indicators; reluctance of elites to address the issue insulates citizens from discursive signals in the public sphere); and self-interest, justice and humanity's common fate (powerful and prosaic forces, from the fossil fuel industry to enjoyment of carbon-intensive lifestyles, generate pressure to maintain the status quo) (Moser 2010: 33–6). Taken together, these factors suggest that climate change 'is difficult to perceive and understand for most lay audiences' and are 'easily trumped by more direct experiences' (p. 36).

Paradoxically perhaps, the more difficult climate change appears to be to understand and care about, the more seductive becomes the idea that the missing link in generating engagement is (more) information and knowledge. This is likely to be especially true for those who are already both well-informed and deeply concerned about the problem (e.g. climate scientists, environmental advocates, sympathetic journalists, etc.): 'if only they understood how severe the problem is', runs the familiar refrain, 'if only we could explain the science more clearly, train to be better communicators, become more media savvy, get better press coverage. . . . Why are they not listening? Why is no one doing anything?' (Dilling and Moser 2007: 3). The dogged persistence of this view means that the institutional response to the failure of information campaigns to produce deeper levels of engagement is often more of the same (Whitmarsh *et al.* 2013: 11). 'When people aren't convinced by hearing the scientific facts of climate change', quips environmental psychologist and consultant Per Espen Stoknes, 'then the facts have been repeated and multiplied. Or shouted in a louder voice. Or with more pictures of drowning polar bears, still bleaker facts, even more studies' (2015: xix). A more radical explanation observes that these failures are, in fact, entirely functional for dominant economic and political interests which are threatened by real mitigation: Sisyphean obsessions with the failure of information campaigns to motivate behavioural change (which actually do little but download culpability and responsibility for the problem to individuals) provide excellent cover for those committed to business-as-usual (Maniates 2001; Shove 2010; Webb 2012).

A growing number of academics and practitioners (e.g. Hulme 2009; Marshall 2014; Moser 2010; Norgaard 2011; Potter and Oster 2008; Stoknes 2015) have argued that this linear pathway of communicative engagement – from information provision to understanding to concern to action – is simplistic and naïve with respect to the range of psychological, cultural, social, institutional and structural variables which shape how (and if) people engage with climate change. Susan Owens, for example, points out that the basic 'myth' of an information deficit model – 'if people had more information about, and better understanding of, environmental risks, they would become more virtuous or might at least accept with better grace otherwise unpalatable policies' – is contradicted by 'a substantial body of social-scientific research [which] suggests that, while greater knowledge may be worthwhile in its own right, barriers to action do not lie primarily in a lack of information or understanding. More important mediating factors are the framing of problems, social and political context, and personal and institutional constraints' (2000: 1142–3).

In a widely cited study based on focus group research with British citizens, Irene Lorenzoni *et al.* (2007) inventory the principal barriers that compromise public engagement with climate change. Without question, a lack of basic knowledge (and, by implication, inadequate access and exposure to credible information about climate change) contribute to key engagement barriers, including uncertainty about climate science; perception of climate impacts as distant in time and space; and distrust of scientific, government and media information sources. However, equally prominent are a range of additional barriers, at both the individual and social levels, which have less to do with information and knowledge *per se* and more to do with broader conditions (both material and cultural) which shape how people make sense of information, respond to it emotionally, and use it to inform their behavioural choices (also see Dilling and Moser 2007). Barriers at the individual level include: externalization of responsibility to others, faith in technological solutions, assessment of other issues (e.g. personal finances, employment, local environmental problems, etc.) as more salient and pressing, reluctance to adopt lifestyle changes, cynicism about the possibility of mitigation and feelings of personal helplessness to make any meaningful contribution to solving a global problem. At the level of society, people expressed a profound lack of trust in the willingness and capacity of governments and businesses to take any action, as well as scepticism about the motives and intentions of the general public, leading to concerns about the free-rider effect (i.e. why should I do anything if others do nothing). Other significant barriers to engagement and behavioural change include a systematic lack of enabling infrastructure and mechanisms (e.g. mass transit, renewable energy, affordable organic food, etc.) and the ubiquitous inertia of carbon-intensive social norms and expectations, often ingrained in the form of largely unconscious habits which are not deliberately 'chosen' in any rational fashion (2007: 446–53). Many of the most powerful barriers to engagement, in other words, are likely to be impervious to the effect of more and better information about climate change.

Confronted by the limits of the information deficit approach, many in the field of environmental communication have called for a more creative, ambitious and even experimental perspective on climate change communication which explicitly strives to produce deeper forms of engagement that involve a composite of cognitive, emotional/affective and behavioural factors (Nerlich *et al.*

2010; Ockwell *et al.* 2009; Whitmarsh *et al.* 2013). 'It is not enough for people to *know* about climate change in order to be engaged', assert Lorenzoni and her colleagues. 'They also need to *care* about it, be *motivated* and *able to take action*' (2007: 446; emphasis added). Moser, for example, distinguishes three broad 'purposes' for such communication: first, '*to inform and educate individuals about climate change,* including the science, causes, potential impacts and possible solutions'; second, '*to achieve some type and level of social engagement and action* . . . behavioral (consumption-related action) and/or political (civic action)'; and, third, 'trying to foster not just political action or context-specific behavior modification, but *to bring about changes in social norms and cultural values* that act more broadly' (2010: 38; emphasis in original). These objectives are synergistic and interdependent, with success in any one area conditional upon progress in the others. Cognitive engagement with the 'facts' of climate science, for example, is highly dependent upon the social norms and cultural values that frame the context in which such information is encountered (Kahan *et al.* 2011). Joining colleagues from the humanities and natural sciences, Nisbet *et al.* make a compelling case for bringing together four disciplinary cultures – environmental sciences, philosophy and religion, social sciences, creative arts and professions – to develop 'a new communication infrastructure, in which the public is (1) *empowered* to learn about the scientific and social dimensions of climate change, (2) *inspired* to take personal responsibility, (3) able to constructively *deliberate* and meaningfully *participate*, and (4) emotionally and creatively engaged in *personal change* and *collective action*' (Nisbet *et al.* 2010: 329; emphasis added).

Such ambitions sit uneasily with the normative commitment of journalism to a 'monitorial' role (see the discussion in the Introduction and Chapter One) which prioritizes the provision of information to audiences, but eschews forays into advocacy and mobilization as lying beyond its mandate. For the most part, conventional journalistic practices and principles rest upon a belief in the unalloyed power of (good) information to enlighten, engage and motivate and the assumption that people are willing participants in the transfer of such information – that is, they have a genuine interest in learning about issues that affect them, either personally or as part of a larger community. News media may be justifiably criticized for their failure to accurately represent different aspects of climate change or the sheer paucity of coverage given the severity, magnitude and urgency of the threat. But can we hold journalism accountable for enhancing emotional and behavioural engagement with climate change? Should we measure the quality and success of news coverage by its capacity to generate rising levels of interest, alarm *and* action? Do aspirations to empower, inspire and mobilize overstep the boundaries of the Fourth Estate?

In the inaugural essay of the journal *Environmental Communication*, and as noted in Chapter One, scholar and activist Robert Cox invited his peers to reflect upon whether environmental communication ought to consider itself a 'crisis discipline' with a corresponding ethical duty to 'enhance the ability of society to *respond appropriately* to environmental signals relevant to the well-being of both human civilization and natural biological systems' (2007: 15) (emphasis added). Does climate journalism share a similar ethical duty? This emphasis upon response does not prescribe a singular course of action, but it does insist that communication about climate crisis must be more than simply informative; it must also strive to enhance the personal and collective agency of people and communities to act in the face of crisis. This requires moving beyond the provision of information to exploring the social, cultural and political conditions which affect *how* people understand and engage with different types of information. Why and how does information become meaningful, useful, engaging, compelling and empowering – a means of enlightenment, a tool for problem-solving, a spur to action? These are critical questions not just for activists or advocates but for anyone engaged in communicating with the public about climate change, including journalists.

Writing in response to Cox's assertion that environmental communication is a crisis discipline, Susan Senecah noted that 'the crisis is not that environmental situations have become dire, although they certainly have. The crisis is that people have become increasingly cynical about all levels of civic life to address them' (2007: 27). This fundamental (re)definition of climate crisis as communicative and democratic in nature (rather than simply ecological) should inform the vision, ethics, objectives and practices of climate journalism, and reorient it in favour of a 'facilitative' and especially 'radical' role. At its core, this conceptual shift both elevates and complicates considerations of engagement and efficacy, demanding that journalism (like other forms of environmental communication) be mindful of the multiple barriers that hinder engagement with climate change, attentive to how different approaches to news and communication can either reinforce or mitigate those barriers and willing to prioritize and experiment with interventions which enhance political efficacy. In the pages that follow, I offer four deliberately provocative suggestions which could help shift climate journalism in this direction: prioritize audiences who are most likely to engage with news about climate change (rather than a 'one-size-fits-all' approach addressed to the general public); make greater use of a climate justice frame which spotlights the ethical, political and normative dimensions of climate change; foreground and 'activate' cultural values which are most likely to promote pro-environmental attitudes, beliefs and behaviour; and cultivate social norms of civic engagement and political efficacy with greater attention to the stories, experiences and emotions of people and communities who are working together to address climate change. Drawn from research in environmental communication (and related disciplines), these four strategies do challenge conventional journalistic norms and practices, but they also keep faith with journalism's basic democratic mandate to enhance public engagement with the most pressing contemporary issues.

Engaging publics: The *alarmed* and the *concerned*

How one conceives of an audience shapes how one communicates with it: In the case of climate change, assumptions about audience needs, interests, scientific and political literacy, beliefs and values inform key decisions about news and editorial content, including areas of emphasis (and avoidance), framing, sourcing, diversity of argument, level of complexity and analysis, tone and style. There are many different ways of conceiving the primary audience for climate journalism. From the perspective of the information deficit approach and the monitorial role, the audience is predominantly framed in an abstract, universalized and rational fashion: everyone is equally interested in learning about climate change, and consequently, the principal objective of climate journalism should be to provide neutral, objective and balanced information that all citizens, irrespective of their particular social locations, economic interests or political sympathies, can draw upon in understanding the issue.

As most recognize, this idealized image of a homogeneous, rational public stands in stark contradiction to actually existing publics, especially in the Anglo-American world, in which there are deep divisions about climate change (McCright and Dunlap 2011). In the United States and, to a lesser extent Australia and Canada, conservative segments of the population have grown suspicious of climate science and deeply hostile to the prospect of more aggressive climate policies. These trends have been fuelled and exacerbated by the ideological balkanization of core sectors of the news media, including the entrenchment of a largely self-contained (and self-referential) right-wing 'echo-chamber' (Jamieson and Capella 2008) in which cues from conservative political and media elites reinforce 'scepticism' about climate science and politics. These tendencies have been intensified by systematic misinformation campaigns about climate science and climate policy, well-financed by wealthy individuals and the fossil fuel industry and implemented by a coordinated network of public relations forms, conservative think tanks and a handful of high-profile 'scientists' (Hoggan with Littlemore 2009; McCright and Dunlap 2010; Oreskes and Conway 2010). As sociologists Aaron McCright and Riley Dunlap argue, these conditions have severely degraded the public sphere, raising doubts about the possibility of civil dialogue:

> New information on climate change (e.g., an IPCC report) is thus unlikely to reduce the political divide. Instead, citizens' political orientations filter such learning opportunities in ways that magnify this divide. Political elites selectively interpret or ignore new climate change studies and news stories to promote their political agendas. Citizens, in turn, listen to their favored elites and media sources where global warming information is framed in a manner consistent with their pre-existing beliefs on the issue.
>
> (McCright and Dunlap 2011: 171)

Climate change has become yet another culture wars battleground in which ideological beliefs, cultural values and partisan affiliation operate as perceptual, cognitive and affective filters that determine how and even if people engage with news and information about climate change (Kahan *et al.* 2011; Wood and Vedlitz 2007).

How should climate journalism respond? Some argue that polarization demands communication strategies through which different groups could find some common ground (e.g. Corner 2013; Marshall 2014). Many have expressed optimism, for example, that prioritizing issues of energy use and policy can successfully avoid the suspicion and hostility triggered by mention of climate change (Leiserowitz *et al.* 2014a). Others assert that the creative use of specific framing strategies, such as accenting the public health, economic development or environmental stewardship dimensions of climate change, can successfully engage constituencies which have otherwise proven difficult to reach (Nisbet 2009). These approaches are consistent with a facilitative role for journalism and hold considerable promise in terms of both improving the quality of public discourse and exploring possibilities for bipartisan compromise and consensus. They deserve attention from climate journalism and are discussed further in Chapter Five.

However, prioritizing audiences which are uninformed, apathetic or sceptical also risks reproducing the flawed and impotent logic of the information deficit approach and, more important, can divert scarce communications resources away from those most likely to become more fully engaged with climate change. Most opinion polling on climate issues, for example, portrays the public as divided into two opposing camps: those who accept the science and/or support action versus those who do not. This representation implicitly elevates the latter group as the principal target for communicative intervention, suggesting that shifting their opinions (and building consensus, or at least stronger majorities, in support of action) is the prerequisite for social change. Unfortunately, however, this logic neglects those who are actually the most promising constituencies for climate journalism.

Since 2008, researchers from Yale and George Mason Universities have been conducting detailed surveys of U.S. public opinion about climate change. They divide the general population into six distinct segments – ranging from the *alarmed* to the *dismissive* (Figure 2.1) – which each respond to information about climate change in different ways based upon varying levels of awareness, concern and engagement. Good climate change communication, they argue, must take into account the different perspectives, values and beliefs held by each of these publics. 'Messages are unlikely to be effective if a diverse population is treated as a homogenous mass' (Roser-Renouf *et al.* 2015: 368).

The *alarmed* and the *concerned* possess high levels of issue involvement and are predisposed to accept and carefully process information about climate change. The *alarmed* are certain that anthropogenic global warming is happening, believe that people (including in the United States) are currently being harmed by it and worry that their families and future generations are at risk. Three quarters see climate change as potentially solvable (Roser-Renouf *et al.* 2015: 372–3). The informational needs and media use of this segment are especially striking:

> They are very attentive to global warming news, compared to other segments: 55 percent report paying 'a lot' of attention to news stories about

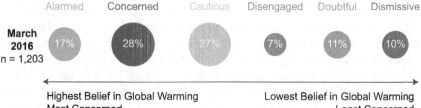

Figure 2.1 Six segments of the U.S. public ('Six Americas') based on beliefs, attitudes and behaviours about climate change.

Source: Roser-Renouf, C., Maibach, E., Leiserowitz, A., & Rosenthal, S. (2016) *Global Warming's Six Americas and the Election, 2016*, New Haven, CT: Yale University and George Mason University. Yale Program on Climate Change Communication. Reproduced with permission from the authors.

global warming, more than four times as high a proportion as any other segment. Almost 80 percent of the *Alarmed* follow environmental news, compared to a national average of 38 percent, and over half say they pay 'a lot' of attention to global warming information.

(p. 373)

Levels of involvement for the *concerned* are not quite as high but are significantly greater than in other segments. A substantial majority see global warming as a risk to their families and future generations, and more than two thirds see climate change as a problem that humans could solve. Close to three quarters pay at least 'some' attention to information about global warming and they are 'more likely than average to follow environmental news' (pp. 373–4). Both the *alarmed* and the *concerned* are 'receptive to messages with a great deal of information and complexity, including relatively high-level science and policy content', and 'because messages . . . will likely be processed effortfully, message content is more likely to be remembered, and behavioural changes are more likely' (pp. 374–5).

Contrast these characteristics to the 'low-involvement' segments, the *cautious* and the *disengaged*, in which 70 to 80 percent report paying 'little or no attention' to global warming information. More than three quarters of the *disengaged* and close to half of the *cautious* say they 'have difficulty understanding news reports about global warming'. In terms of their affective dispositions, close to 60 percent of the *disengaged* and 40 percent of the *cautious* state that they 'don't like to read or hear anything about global warming' (pp. 376–9). The final two segments – the *doubtful* and the *dismissive* – report being somewhat more engaged, but they also bring a deep-seated scepticism to encounters with news about climate change. The *doubtful* 'take a dim view of the notions that humans have caused climate change or can solve it', 'few think that scientists agree climate change is happening', and they pay minimal attention to information about global warming or environmental news. The *dismissive*, who tend to attract a lot of concern and attention in public debates about climate change given the high profile their views receive in conservative media, are 'the most certain that climate change is not happening', 'highly confident in their views' and 'the most likely of any segment to say that they do not need any more information to make up their mind on the topic' (Roser-Renouf *et al.* 2015: 380–1). The simple fact is that these four segments pay little attention to news about climate change and, when they do, they have a strong disposition to dismiss information that clashes with their existing views.

If we accept that the alarmed and the concerned are the primary constituency for climate journalism, what are the most promising strategies for increasing their levels of engagement with climate change? Broadly speaking, the provision of more information about climate science is unlikely to have much impact because these groups are already convinced of the reality, danger and human-caused nature of climate change. Instead, they are far more interested in news about solutions. Asked what question they would most like to pose to an expert on climate change, the alarmed and the concerned overwhelmingly prioritized information about *actions* and, in particular, information about political actions and policy-making (rather than consumer or lifestyle change; see Figure 2.2 and Leiserowitz *et al.*

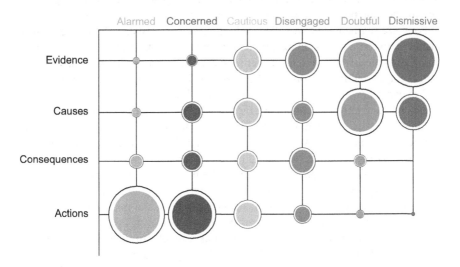

Figure 2.2 Nature of the one question Americans would most like to pose to a climate scientist.

Source: Roser-Renouf, C., Stenhouse, N., Rolfe-Redding, J., Maibach, E. and Leiserowitz, A. (2015) 'Engaging diverse audiences with climate change: Message strategies for global warming's six Americas', in A. Hansen and R. Cox (eds), *Routledge Handbook of Environment and Communication*, New York: Routledge, p. 371. Reproduced with permission from the authors.

2011: 15). 'The challenge with the high involvement segments is motivating them to take action, *particularly political action and opinion leadership*: Even among the *Alarmed*, political actions are not the norm' (Roser-Renouf *et al.* 2015: 374; emphasis added). Both groups have 'high levels of concern about climate change, but lower levels of efficacy with regard to solving it; hence, communicators may wish to focus on *building efficacy* to complement the groups' high risk perceptions to motivate them to take action' (p. 375; emphasis added).

Aware, concerned and motivated to take action, these high-involvement segments – which compose more than a third of the population – represent a potential tipping point in intensifying public engagement. Three quarters of the alarmed and one third of the concerned, for example, say they are willing to join a political campaign to convince governments to take action on climate change compared with much smaller proportions for the other segments (Leiserowitz *et al.* 2014b: 20). Both groups possess a strong belief in the *potential* for collective political action to stimulate more aggressive government action on climate change, but they also report pessimism about their *actual* political efficacy, assessing their own collective influence upon policy agendas as much weaker than other groups such as large campaign contributors and the fossil fuel industry (Leiserowitz *et al.* 2013: 26–8). They believe in the necessity of collective action and are (hypothetically) willing to become politically engaged, but they are also uncertain about their capacity to make social change (Cross *et al.* 2015). The key to shifting these groups up the ladder of engagement, in other words, involves giving them a much stronger practical understanding of the possibilities and positive impacts of active participation in climate politics.

For these segments, the so-called 'values-action' gap (Blake 1999) – the contradiction between a professed commitment to environmental values and reluctance/ inability to modify behaviour – is better understood as a gap between an abstract desire for collective action and the practical knowledge, behavioural scripts and experiences of solidarity that transform political aspirations into real civic engagement. Cultivating political efficacy requires a more equitable balancing of attention between the dominant institutions and events of climate science and politics (e.g. the Intergovernmental Panel on Climate Change, annual global negotiations, etc.) and the vast but all too often invisible multiplicity of organizations, communities and practices through which people are organizing to address climate change (Gunster 2011, 2012). It depends upon giving audiences a much broader, deeper and more grounded understanding of the many different ways that citizens can participate (and are *already* participating), both formally and informally, in the political process, as well as a much stronger appreciation for the value and impacts of such participation. It requires a greater focus upon the intersection between citizen engagement, political institutions, public policy and the implementation of collective solutions to climate change. Above all, perhaps, it means resisting the temptation to define the primary audience for climate journalism as those who are indifferent, apathetic, poorly informed or uninterested in climate change and leave 'preaching to the choir' to advocates and special interests. Instead, climate journalism might productively reorient itself to serving the needs and interests of those most likely to engage with it, and it might imagine and address its audience as citizens who are aware and concerned about climate change and who seek credible information about its causes, consequences and solutions but, above all, are looking for ideas, knowledge, inspiration and stories through which to envision themselves as agents of change.

Framing climate change: Efficacy, politics, justice

Framing is arguably the most common conceptual tool applied to climate change communication, both in terms of analyzing news coverage and other forms of media (e.g. Feldman *et al.* 2015; Hart and Feldman 2014; Olausson 2009; O'Neill *et al.* 2015) as well as providing strategic advice to practitioners about how to communicate more effectively with a diverse range of audiences (e.g. ecoAmerica 2009; Lakoff 2010; Nisbet 2009; Topos Partnership 2009). 'To frame,' explains Robert Entman, 'is to *select some aspects of a perceived reality and make them more salient in a communicating text, in such a way as to promote a particular problem definition, causal interpretation, moral evaluation and/or treatment recommendation* for the item described' (1993: 52; emphasis in original). While the bulk of scholarly analysis in this area has focused upon how the *science* of climate change has been framed, there is increasing awareness that the question of how climate *politics* is framed may actually have greater impacts upon levels of civic engagement (Cross *et al.* 2015; Gunster 2012). Such engagement, argues Anabela Carvalho, is inevitably informed by the accounts of climate politics one finds in news media which 'are the main arenas for citizens' understanding of political struggles in our times' (2010: 174). From the perspective of cultivating political agency, dominant frames of climate politics as the exclusive preserve of politicians, bureaucrats, technical experts and corporate leaders leaves little space for more active conceptions of citizenship. 'As political agents with positions, ideas, and proposals for addressing climate change and other public matters, citizens have been largely left out of the media(ted) discourses. These exclusionary constructions do not recognize citizens as worthy speakers on the substance of collective problems and do not cultivate a proactive political identity' (Carvalho 2010: 175).

In a comprehensive study of U.S. network news coverage of climate change between 2005 and 2011, P. Sol Hart and Lauren Feldman confirmed that citizen efficacy (i.e. the ability of individuals to participate in climate politics) does not feature prominently in the news mix:

> News broadcasts devoted minimal attention to external efficacy, or the responsiveness of government officials to public calls for action, despite extensive coverage of proposed governmental action to reduce climate change. Thus, government action was typically portrayed as divorced from public opinion rather than as a response to calls for action by individuals and advocacy groups. Only a small minority of broadcasts described personal or political actions that individuals can voluntarily take to address climate change, and

an even smaller number of broadcasts emphasized the self-efficacy related to these questions.

(2014: 341)

Discussion and debate about climate policy, in other words, is largely presented as a political spectacle: citizens may passively observe from afar, but real control over policy lies within institutions and processes which are primarily responsive to powerful special interests. The ubiquity of such accounts nurtures the cynicism and alienation which, as previously noted, is one of the primary barriers to political engagement among those concerned with climate change (Cross *et al.* 2015). A companion study of U.S. print media found a similar marginalization of agency: 'there is little readers can glean from these stories about what actions they, as individuals, can take personally or politically to address climate change.' Framing of action predominantly 'stressed political conflict and strategy, which . . . likely portrays climate change as a seemingly intractable problem whose potential solutions are mired in the competition and self-interested motivations of dueling political elites' (Feldman *et al.* 2015: 13). In the absence of concrete, tangible representations of how citizen action can influence climate politics, news about climate change often does little more than intensify feelings of anxiety, fear, alienation and helplessness which are ultimately toxic to engagement.

Does a political/conflict frame inevitably undermine efficacy? Elsewhere I compared corporate with alternative media coverage of the 2009 Copenhagen summit and identified two significantly different frames for climate politics. At one level, these findings were broadly similar to those previously described in this chapter in which corporate media portrayed the political sphere as little more than a 'space of endless bickering and intractable gridlock, with participants both unwilling and unable to deliver any effective action on climate change' (Gunster 2011: 490). Alternative media, however, provided a much richer, more complex and engaging account of climate politics, both offering and demanding:

a far more active and engaged political sensibility in which outrage with existing institutions was cause for action and not despair. . . . Diagnostic assessments of the limits of conventional politics and existing institutions inspired calls for *more* rather than less political engagement, a demand that citizens actively confront those with power and influence rather than abandon the political field to their control.

(Gunster 2011: 492)

Criticism of the spectacular failures of dominant institutions and processes was balanced with coverage of effective climate policies and actions which have been undertaken by governments, attention to the desire of many citizens to trade hyper-consumerism for more sustainable communities and, above all, attention to the many forms of climate activism through which 'ordinary' people were working together to force their governments to be more responsive to popular concern about climate change (Gunster 2011: 490–8).

Critical analysis and application of political frames to news about climate change must become far more attuned to the different and often contradictory ways in which such frames can *enable* (as well as constrain) public engagement. In a recent study of news coverage (O'Neill *et al.* 2015), for example, researchers developed multiple frames for several issue areas – science, economy, morality/ethics – in recognition of the complex and often contradictory manner in which they are framed: science frames may emphasize certainty or uncertainty, economic frames may emphasize the costs/risks of action or inaction, and invocations of morality may be used to support or oppose climate action. In the case of politics, however, only a single frame of 'political or ideological struggle' was used, defined as 'a conflict over the way the world should work: over solutions or strategy to address climate change . . . a battle for power (for example, between nations or personalities)' (O'Neill *et al.* 2015: 381). No allowance was made for the crucial difference between a frame which defines politics largely as an elite enterprise and one which adopts a more radical, participatory vision of politics as a process of citizen engagement and activism. In a similar fashion, Hart and Feldman rely upon a single 'conflict/strategy' frame which is positioned as corrosive of political efficacy: 'the emphasis on political conflict in news coverage is likely to give the impression that the government will be unresponsive to calls for action' (2014: 344). Essentialized definitions of political/conflict frames as inherently alienating not only risk replicating the one-dimensional, monochromatic accounts of climate politics delivered through corporate media, but also they ignore how stories of political struggle are among the most potent frames in getting people excited about political action. This logic is explored at greater length in Chapter Five, but for now it is enough to assert the need to restore a more supple, dynamic and (potentially) optimistic approach to climate politics which recognizes the positive, generative role that conflict, politicization and even polarization can play in strengthening engagement with climate change.

Frames of environmental justice, which initially emerged out of the experiences of people of colour in the United States, provide a far more robust template for framing the politics of climate change. Drawing upon the sociological concept of 'collective action frames' (Benford and Snow 2000), Dorceta Taylor identifies three primary attributes of the environmental justice frame: *injustice*, which foregrounds a 'moral indignation [that] is more than a straightforward cognitive or intellectual judgment about equity or justice; it is a "hot" cognition – one that is emotionally charged'; *agency*, which 'refers to individual and group efficacy' such that 'those exercising agency feel they can alter conditions and policies'; and *identity*, 'the process of defining the "we" or "us" – usually in opposition to "they" or "them"' (2000: 511). The environmental justice frame evolved to make sense of shared experiences of environmental racism, built upon existing traditions of agency and activism cultivated through the civil rights movement, and helped push environmentalism beyond a (pristine) nature/wilderness focus which often prioritized places over people. Ultimately, the frame is simultaneously transformative of both consciousness and behaviour, enabling a form of 'cognitive liberation' through which individuals recognized the illegitimacy of dominant social and political institutions, accepted the possibility of social change and started to imagine and exercise new forms of political efficacy (Taylor 2000: 520).

Attention to climate justice has grown rapidly over the past decade (Bond 2012; Klein 2014; Shue 2015), driven in large part by the extraordinary coalescence of a wide range of organizations and communities (e.g. Indigenous people, workers, peasants, local communities, etc.) around a shared experience and (re)framing of climate change as injustice. Where a scientific frame suggests the key to understanding climate change is its anthropogenic origins in greenhouse gas emissions and the catastrophic impacts if such emissions continue to rise, a climate justice frame insists that the most important thing to know about the problem is the highly unequal and grossly unfair distribution of risks, responsibility and benefits. Simply put, those who are least responsible for causing climate change will suffer the most harm from its impacts, while those who bear the most responsibility will not only suffer the least but also are, in fact, the principal beneficiaries of fossil fuel use. To put this another way, climate justice defines the root cause of climate change *not* as emissions but as *inequality* – a pervasive, structural inequality that systematically divorces responsibility from accountability and thereby violates the core normative principle of distributive justice. Adding insult to injury, those most responsible for this injustice continue to dominate discussions of mitigation, leaving those most likely to be victimized by climate change with little opportunity to participate or even influence decisions which will affect their fate. Finally, the most severely affected are also the least likely to possess adequate resources for adaptation, leaving them both vulnerable to climate impacts and dependent upon the assistance of others.

Framing analysis typically subsumes these considerations within a moral or ethical frame which is differentiated from political frames (e.g. Hart and Feldman 2014; Nisbet 2009; O'Neill *et al.* 2015). These issue-based framing silos, however, cut against the grain of how those involved in the movement actually define climate justice: 'for many of the groups and networks that participated in [the Climate Justice Action network] the broad position underlying the use of the term is the *politicisation of climate change* – understanding that it results from our current and historical social relations, and that in order to address it we need fundamental changes to our economic and political systems' (Building Bridges Collective 2010: 27; emphasis added). Unlike conventional approaches to media or issue framing (which rest primarily upon a taxonomy of division that identifies the inclusion of particular characteristics and the exclusion of others), collective action frames embody a more dynamic approach which knits together a variety of thematic elements to broaden and amplify the salience, credibility and resonance of particular messages. The climate justice frame depends upon expanding and strengthening the conceptual, affective and behavioural affinities between ethics, morality and politics which are otherwise impoverished when these different frames are kept separate. Ethical considerations of climate justice, for example, remain abstract and idealized in the absence of more substantive accounts of the movements which are actively fighting to achieve these objectives.

A full exploration of how the principles of climate justice could be applied to climate journalism lies beyond the scope of this chapter (though see useful discussions in Hackett *et al.* 2013; Schmidt and Schäfer 2015). It is clear, however, that foregrounding questions of climate justice could motivate (and empower) journalists to explore a broader range of questions and perspectives about climate change and provide an abundance of resources for audiences to conceptualize the issue in novel ways. Consider, for example, the simple matter of how statistics about emissions are presented: a justice frame would mandate (or at least encourage) bringing social, historical and ethical perspectives to these metrics in order to illuminate patterns of inequality and facilitate deeper understanding of questions of responsibility. Foregrounding historical (rather than annual) accumulation data could enhance greater awareness and acceptance of the principle of common but differentiated responsibility, a cornerstone of the United Nations Framework Convention on Climate Change. Likewise, the distinction between subsistence and luxury emissions – a simple and intuitive yet compelling idea – could promote the acceptance of steeper emissions reductions in wealthier nations, build support for climate policies targeted to particular forms of consumption (e.g. flying) and ensure that the effects of carbon pricing do not fall disproportionately upon those least able to bear the burden. Closer attention to emissions accounting regimes, which 'book' the carbon footprint of commodities to the country in which they are produced, not consumed, could inspire greater awareness of the problem of emissions offloading (and temper sanctimony about ostensible reductions which have actually just been transferred to others). More frequent use of per capita calculations across all metrics would enhance the comparability of different entities (e.g. countries, economic sectors, income levels, etc.) and implicitly reinforce basic principles of equality by persistently highlighting existing levels of inequality. These are but a few examples of how applying a climate justice frame to emissions data could introduce a much broader range of moral, ethical and political perspectives on questions of responsibility and action.

More broadly, though, the climate justice frame directly confronts indifference and cynicism, two of the most persistent barriers to public engagement. Just as environmental justice activists forced the environmental movement to expand its horizon beyond questions of wilderness, climate justice forces us to confront the human impacts of climate change, complementing images of polar bears and melting ice caps with people who can speak with authority, eloquence and passion about the devastating implications for the places and communities they love. The most compelling and persuasive climate change communicators are often not scientists, politicians or celebrities: instead, they are real people telling stories about their experiences and lives. These stories have the capacity to connect with us, often at a deeply emotional level, making us care about the people and places they describe in a more powerful and lasting fashion than any fact or figure. Even more important than (re)defining climate change as injustice, a climate justice frame challenges the cynical acceptance of the status quo as circumscribing the totality of politics. Stories of people coming together to fight injustice are powerful tools in the generation of political efficacy. The accumulation of such stories legitimates and naturalizes political engagement as an easy, practical and effective response to concern about climate change. Attention to grassroots forms of civic engagement is likely to be particularly enlightening and appealing to the primary audiences for climate journalism who have grown sceptical of individualized, lifestyle-based behavioural change and deeply cynical about conventional forms of climate politics.

Shifting values: From self-interest to common cause

In a widely cited 2005 study of the U.S. public's risk perceptions of climate change, Anthony Leiserowitz concludes that 'most Americans lack vivid, concrete, and personally relevant affective images of climate change', instead associating its impacts with 'places or natural ecosystems distant from the[ir] everyday experience' (2005: 1438). Invoking former U.S. Senator Tip O'Neill's famous maxim that 'all politics is local', he suggests that 'climate change is unlikely to become a high-priority national issue until Americans consider themselves *personally* at risk' (2005: 1438; emphasis added). According to this logic, then, a key challenge for communicators is to provide individuals with a more concrete, immediate and visceral sense of how climate change will directly affect themselves, their communities and their local environments. Insofar as these strategies increase the salience of climate change by localizing information to fit the knowledge and experiences of particular audiences (and challenge the 'optimism bias' which leads many to underestimate the local risks of climate change), they can significantly improve the quality and impact of advocacy campaigns and climate journalism (Rootes 2007; Shaw *et al.* 2009).

An excessive focus upon the individualization of risk, however, can also incubate the more insidious logic that plots a person's level of engagement purely as a function of self-interest, suggesting that the most effective (and perhaps the only) way to motivate someone to care is to persuade them that climate change endangers their individual safety, security and well-being. People are presumed (and encouraged) to modulate their level of interest and concern based upon the extent to which they feel personally threatened. And climate change communicators (including journalists) are locked into a communications strategy which pursues intelligibility and resonance by boiling down the complexities of climate change to images or facts which can convey the immediacy of the threat in an emotional, visceral manner. Communication about behavioural change and climate policy is often subject to an inverted but similar promotional calculus: support for such measures depends upon the extent to which individuals come to realize (and appreciate) the benefits that will flow directly to them.

In earlier work, I analyzed public discourse around carbon taxes as exemplary of this logic (Gunster 2010). Most advocates suggest that public support for a carbon tax hinges upon designing and framing it as a revenue-neutral form of tax-shifting (which provides financial benefits to consumers) rather than, for example, as a means of funding public infrastructure projects such as expanding mass transit or renewable energy. The persistence of this neoliberal perspective is quite striking given the fact that opinion polls consistently show that support for carbon taxes actually rises when such revenues are *collectively* allocated to investments in sustainability rather than dispersed to individuals through tax-shifting (e.g. Amdur *et al.* 2014; Leiserowitz *et al.* 2013: 18–21; Pratt 2015). But in a world dominated by consumer culture and neoliberal political discourse, the hegemony of individualization (Maniates 2001) and self-interest is difficult to resist, even for those who may not subscribe to those values themselves:

> What's in it for me? Forty years of neoliberal hegemony have ravaged and stunted the radical political imaginary such that we twist ourselves into ideological knots trying to always answer this question in the most conservative of terms. . . . In bringing our political ambitions and discourse down to this level, though, we call into being the fictional public that we fear the most: selfish, apathetic, and motivated by nothing other than the utilitarian calculus of financial cost and benefit.
>
> (Gunster 2010: 208)

The materialistic, individualistic values of neoliberal political culture, however, are not necessarily reflective of the values of the broader public, and more important, they actually stand in stark contrast to the values of those most likely to engage with climate journalism. The alarmed, for example, are far more likely to hold egalitarian and communitarian values, and are much less individualistic than the national average (Roser-Renouf *et al.* 2015: 8). Building communications strategies around self-interest may, in other words, be missing the key drivers that actually lead many to pay attention and care about climate change.

Decades of research into risk perception and environmental behaviour affirm the strong affinities between what one might generally describe as altruistic values – care and concern for the well-being of others (both human and non-human) – and constructive engagement with environmental issues. Values play a critical role in determining how and even if individuals perceive, process and assess different types of risks (Douglas and Wildavsky 1982; Kahan *et al.* 2011). Often described as 'confirmation bias' (Nickerson 1998) or 'motivated reasoning' (Kunda 1990), an individual's level of engagement with information depends, in part, upon the extent to which it confirms or challenges his or her existing values, beliefs and worldview. Those with greater commitments to individualistic values which emphasize self-interest and personal freedom, for example, tend to minimize or ignore environmental risks because they perceive that taking such risks seriously would require a more expansive role for government, greater regulation of markets and business activity and constraints upon personal consumption. Conversely, those who prioritize egalitarian and communitarian values are more likely to accept and attend to such risks for precisely the same reasons, because mitigating such risks requires collective action, stronger governments and constraining corporate power, all of which are consistent with the valorization of equality and community well-being.

A similar logic holds true with respect to the adoption of environmentally responsible behaviour and support for environmental policy. Human behaviour is notoriously complex, and it is virtually impossible to identify any single set of variables as definitively constitutive of particular behavioural patterns or choices. However, research in environmental behaviour confirms that 'the more strongly individuals subscribe to values beyond their immediate own interests, that is, self-transcendent, prosocial, altruistic or biospheric values, the more likely they are to engage in pro-environmental behaviour' (Steg and Vlek 2009: 311; also see Dietz *et al.* 2005; Jackson 2005). The same holds true with climate policy: a national survey of U.S. citizens, for example, found that '[s]upport for national and international climate policies was strongly associated with pro-egalitarian values, while opposition was associated with anti-egalitarian, pro-individualist and pro-hierarchist values' (Leiserowitz 2006: 63). Value commitments were stronger predictors of risk perception and policy preference than either party affiliation or ideology. According to Stern's (2000) influential 'value-belief-norm' model of environmental behaviour, altruistic values inaugurate, sustain and strengthen a causal chain of moral reasoning that leaves individuals more likely to subscribe to an ecological worldview which connects human well-being to the health and vitality of the natural environment (defined as the 'new environmental paradigm', or NEP), become interested in learning about the (indirect, imperceptible) environmental consequences of human behaviour, take personal responsibility for the broader impacts of their behaviour upon others, and develop strong feelings of moral obligation to engage in environmentally responsible behaviour. Altruistic values ground a normative ethos in which the pursuit of personal gratification and material self-interest must always be measured against the welfare of others and the public good. Conversely, the stronger one identifies with egoistic or self-enhancement values, the less likely one is to engage in 'pro-environmental' forms of reasoning, affect and behaviour which embody opposing values.

Experimental research into the relationship between empathy and moral reasoning raises challenging questions about whether an ethos of impartial, detached 'objectivity' is appropriate or productive for climate journalism. Environmental psychologist Jaime Berenguer (2010), for example, conducted a study in which the empathy of participants was either primed (by asking them to imagine how the subject of a fictional news item felt about a particular experience) or repressed (by instructing them to remain 'objective' and 'neutral'). Participants were then instructed to review descriptions of several environmental dilemmas (e.g. logging

old-growth forests, building a new landfill), make decisions about them and provide explanations for those decisions. The results showed that priming empathy significantly increased both the volume of arguments and the sophistication
of moral reasoning applied to decision-making about environmental issues. In
a similar vein, Sabine Roeser argues that 'emotions provide us with privileged
epistemic access to moral values, especially when it comes to particular moral
judgments where a complexity of moral considerations needs to be assessed'
(2012: 1035). Thinking of others – and in particular, *feeling for* others – increases
the cognitive engagement that individuals bring to thinking about environmental
issues, and significantly expands the range of moral and ethical considerations that
inform reasoning and decision-making. Invoking emotions associated with empathy, compassion and justice can also help overcome perceptions of helplessness
by grounding agency in the visceral, embodied desire to do something – especially
if such desire can be articulated with stories of like-minded others taking action.

Appeals to self-interest, either by dramatizing climate risks to individuals
and local communities or by accenting the economic benefits of environmental
behaviour and climate policy (e.g. saving money through energy retrofits), are an
obvious and attractive strategy for journalists looking to convince indifferent audiences that they should pay attention to climate change. There are many situations
in which this expository template appears unavoidable, especially in the context
of consumer-driven models of news which elevate the personal needs, desires
and interests of the 'sovereign' reader above all other considerations. However, as
Adam Corner and Alex Randall suggest, 'the tailoring of environmental messages
to promote or enhance self enhancing values could be severely counterproductive
as a strategy for public engagement on climate change' (2011: 1008). Summarizing research on the behavioural impacts of a self-interest frame, Tom Crompton
explains that a strong focus on financial success, for example, is associated with
'lower empathy, more manipulative tendencies, a higher preference for social
inequality and hierarchy, greater prejudice towards people who are different, and
less concern about environmental problems' (2010: 10). Over the longer term, a
self-interest frame reduces the likelihood that people will care and pay attention to
climate news (unless they perceive it as affecting them directly), weakens motivation for behavioural change and perhaps most important, diminishes the appeal of
collective action and political engagement.

Alternatively, journalism that activates and strengthens self-transcending values and cultivates understanding and empathy for the experiences and feelings
of others will generate a more hospitable context for public engagement with
the issue. Counter-intuitively perhaps in a neoliberal culture which presumes that
self-interest is the wellspring for all human thought and action, care for others is
actually a much stronger motivator than care for self in getting people to think
and act on climate change (Howell 2013). And empathy for others is not only
cued through stories and images of suffering and vulnerability but also, more
powerfully, through inspiring accounts of (collective) resistance, struggle and
agency. Heroes, it would seem – especially of the 'everyday' kind – are far more

compelling than either villains or victims in crafting strong narratives that can enhance engagement with climate politics and policies that may otherwise appear distant, abstract and even irrelevant (Cross *et al.* 2015; Jones 2013). Beyond the positive effects such narratives can have upon how audiences engage with individual news items, the cumulative impact of such stories can help disrupt the ubiquitous political culture of neoliberalism by showcasing modes of social and political engagement which embody and make concrete values of equality, solidarity and common cause with others.

Normalizing engagement: Transforming the public's image of itself

Almost two decades ago, the American Geophysical Union commissioned focus group research to explore public sentiments about a range of scientific issues, including climate change. Among the most notable findings was the existence of a strong perception among participants that the origins of many environmental problems can ultimately be traced to an erosion in the moral fibre of the general public:

> When thinking about global warming . . . our respondents typically saw it as being driven by humans who are unwilling to do the right thing, that is a seemingly irreversible deterioration in moral values. What they said, over and over again, was that people have become more self-centered, greedy and materialistic, and as a result, the society is inevitably pushed toward more consumption, which in turn causes more pollution and exacerbates the trend toward global warming.
>
> (Immerwahr 1999: 10)

For a collective action problem such as climate change – in which effective mitigation is widely viewed as dependent upon cooperation, empathy and some measure of self-sacrifice among those with carbon-intensive lifestyles – such bleak perspectives on human nature are especially corrosive for hope and efficacy. A recent report exploring how citizens in British Columbia respond to news about climate change likewise identified a pervasive frustration with the apathy, indifference and materialistic values of 'other people' as a source of cynicism about climate politics. 'The belief that most others do not share one's own values or beliefs can . . . intensify feelings of isolation and helplessness insofar as it becomes increasingly difficult to identify (or even imagine) possibilities for political solidarity and collective action' (Cross *et al.* 2015: 24). Why engage in politics if there is no real hope of ever convincing one's fellow citizens (and, by extension, governments which are presumably accountable to democratic pressure) of the need to take action?

Anke Fischer and her colleagues define such popular perceptions of human nature as 'folk psychology' and argue that it plays a key role in shaping public assessment of environmental behaviour and policy. A study of public opinion in five European countries identified three recurring assumptions about human nature: first, 'humans were described as inherently selfish, considering only their own, and at best, the interests of their own small family group'; second, people were viewed as 'predominantly governed by habit and/or convenience . . . [which] once acquired, were seen as virtually unchangeable and irreversible'; and third, monetary incentives and punishments were reported as 'the only factor that could possibly change people's behaviour' (2011: 1028–9). For many, these psychological attributes are reinforced by reified structures of consumerism, individualism and globalization. 'Overall,' note the research-ers, 'our interviewees painted a rather negative picture of humanity: human ability for cooperation based on insight, morals or voluntary agreements was regarded as limited. Consequently, collective action was not seen as a promis-ing approach to achieve large-scale behavioural change' (Fischer *et al.* 2011: 1033). The origins of this cynicism are complex, but there is little question that the image of the public one encounters in news media bears some of the blame. In a wide-ranging study of media representations of political participa-tion, Justin Lewis *et al.* (2005) found that news media perpetuate a vision of citizens as consumers who think only of their own pleasure and self-interest: 'ordinary' people "speak in individualised and privatised terms, and appear ignorant and childish in their inability to consider the long-term consequences of actions. . . . The implicit ideological message is that the 'common' person is incapable of rational reflection, only living for the moment" (Lewis *et al.*: 81). Such representations nourish deep scepticism about the prospect of get-ting 'others' politically engaged in climate change and, when combined with pessimistic accounts about inaction by governments and politicians, intensify cynicism about climate politics.

Complementing these perceptions are powerful emotional and conversational norms through which considerations of climate change are disciplined, marginal-ized and excluded from everyday life. 'Social norms of attention, conversation and emotion – that is, the social standards of what is 'normal' to think and talk about and feel – are powerful, albeit largely invisible social forces that shape what we actually *do* think and talk about and feel' (Norgaard 2011: 132). Based on exten-sive field work in a Norwegian community, Kari Marie Norgaard documented a ubiquitous 'social organization of denial' in which people who were both aware and concerned about climate change nonetheless avoided thinking or talking about it in any sustained fashion. She traces this dynamic to the fact that thinking about climate change inevitably raises unpleasant emotions – fear, guilt and helplessness – which challenge more positive conceptions of self-identity that people prefer to maintain about themselves. Such emotions are difficult to control once they have arisen; consequently, the most effective strategy to manage them is to regulate (and repress) the thoughts and ideas through which they are triggered.

In addition to the psychological mechanisms through which individuals avoid thinking about troubling topics, Norgaard documented the widespread presence of (informal) social norms which discouraged people from talking about climate change with family and friends, or in institutional settings such as schools, workplaces or even local political events and meetings. And when climate change did arise, a range of discursive strategies were employed to short-circuit more substantive consideration of the topic: irony, humour and cynicism, for example, marginalized efforts to engage in more serious discussion, and the culpability of distant others (especially the United States) was commonly invoked to displace attention from the responsibility of 'little Norway' for this global problem (2011: 174). Norwegians did not lack knowledge or concern about climate change; instead, what was missing were conversational and emotional norms which could have provided cultural support for having open, honest and frank discussions including, most importantly, making it socially acceptable to feel and express the powerful (and troubling) emotions that inevitably accompany any serious talk about the meaning and significance of climate change.

Consideration of behavioural social norms often focuses upon lifestyle and consumption patterns, exploring how norms influence the choices people make and/or the habits that form in areas such as diet, travel and energy use (e.g. Peattie 2010). Commercial media are deeply implicated in promoting carbon-intensive social norms which are a key driver in emissions growth; indeed, 'news' and advertising often operate synergistically in the typical sections of daily newspapers (e.g. cars, home, travel, technology) to motivate and normalize ever-higher levels of consumption. Challenging these norms – or at least exposing the contradiction between such consumption and the need for immediate reductions in emissions – is a critical task for climate journalism. Equally, if not more important, however, is the cultivation of new social norms around the *political* behaviour, conversation, values and emotions associated with climate change. Giving the public a different image of itself – or, more specifically, giving those who are aware and concerned about climate change a much stronger sense of their own shared values and identity, motivation and desire for engagement – is one of the highest and most immediate priorities for climate journalism.

In a fascinating study of the impact of social norms upon environmental behaviour, Vladas Griskevicius and his colleagues (2008) distinguish between injunctive social norms (prescriptions of how people should behave) and descriptive social norms (descriptions of how people are behaving), arguing that the latter are far more effective in motivating and shaping behaviour. Moralized calls for behavioural change often generate unpleasant feelings of guilt, shame and frustration as people feel pressured to do things that will set them apart (and perhaps attract ridicule and censure) from their peers. As Norgaard suggests, such feelings can reinforce social norms which license indifference, dismissal or even hostility to such rhetoric as a strategy of managing the troubling emotions which they often produce.

In contrast, showing the public that others – and preferably others *like them*, with whom they can identify and empathize – are getting engaged may be a far more effective communications strategy, holding greater potential to attract interest, enthusiasm and emulation from sympathetic audiences. The rhetorical appeal of normative but abstract ideals around political engagement – the virtuous practice of a good citizen – is immeasurably enhanced when conjoined with 'everyday' stories about the experiences of 'real' people who are already engaged in this behaviour. Perception of how others 'normally' think, feel and behave are extremely influential in shaping how people respond and engage with different situations, especially those characterized by novelty and/or uncertainty (Bandura 1977). Accordingly, the most effective strategies for public engagement may depend upon the cumulative representation of such behaviour as common, practical, enjoyable and politically effective: in short, as a *normal* response to climate change. Such stories can disrupt debilitating, inaccurate but otherwise ubiquitous stereotypes of the public as apathetic, individualistic and selfish. They challenge the cynical complacency born of the presumption that others do not care, do not share one's values and have no interest in engaging with the issue. Though media, culture and politics may be enamoured with Hobbesian fantasies of moral corruption, there is a different set of 'truths' about human nature. Decades of academic research, notes *Guardian* columnist George Monbiot, have established that humans are 'ultrasocial: possessed of an enhanced capacity for empathy, an unparalleled sensitivity to the needs of others, a unique level of concern about their welfare, and an ability to create moral norms that generalise and enforce these tendencies' (2015).

'If you look at the science about what is happening on earth and aren't pessimistic, you don't understand data. But,' continues Stephen Hawking, 'if you meet the people who are working to restore this earth and the lives of the poor, and you aren't optimistic, you haven't got a pulse' (cited by Moser 2011: xvii). Stories of climate politics and activism can help invigorate a political imaginary stunted by decades of neoliberalism, awakening people to the possibilities and successes of collective action. More than this, however, the prospect of political engagement can transfigure the emotional valences of climate change. The paralytic effects of fear, guilt and helplessness are inversely proportional to perceptions and experiences of efficacy. While those effects – and the consequent social organization of denial – may be an indomitable feature in everyday life, climate politics provides more fertile and expansive terrain through which to think, talk and engage constructively with the emotions associated with climate change. Via the alchemy of political agency, threat becomes challenge, fear becomes anger, helplessness becomes agency and despair becomes hope.

Note

1 My thanks to Chris Russill, Roy Bendor and Adrienne Cossom for their helpful comments on an earlier draft of this chapter.

References

Amdur, D., Rabe, B. and Borick, C. (2014) 'Public views on carbon tax depend on the proposed use of revenue', *Issues in Energy and Environmental Policy* 13 (July): 1–9.

Bandura, A. (1977) *Social Learning Theory*, Englewood Cliffs, NJ: Prentice Hall.

Benford, R. and Snow, D. (2000) 'Framing processes and social movements: An overview and assessment', *Annual Review of Sociology* 26: 611–39.

Berenguer, J. (2010) 'The effect of empathy in environmental moral reasoning', *Environment and Behavior* 42(1) (January): 110–34.

Blake, J. (1999) 'Overcoming the "value-action" gap in environmental policy: Tensions between national policy and local experience', *Local Environment: The International Journal of Justice and Sustainability* 4(3): 257–78.

Bond, P. (2012) *Politics of Climate Justice: Paralysis above, Movement Below*, Scottsville, South Africa: University of Kwazulu-Natal Press.

Building Bridges Collective (2010) *Space for Movement? Reflections from Bolivia on Climate Justice, Social Movements and the State.* Accessed at https://spaceformovement.files.wordpress.com/2010/08/space_for_movement2.pdf.

Carvalho, A. (2010) 'Media(ted) discourses and climate change: A focus on political subjectivity and (dis)engagement', *Wiley Interdisciplinary Reviews: Climate Change* 1 (March–April): 172–9.

Corner, A. (2013) *A New Conversation with the Centre-Right about Climate Change: Values, Frames and Narratives*, London: Climate Outreach and Information Network.

Corner, A. and Randall, A. (2011) 'Selling climate change? The limitations of social marketing as a strategy for climate change public engagement', *Global Environmental Change* 21: 1005–14.

Cox, R. (2007) 'Nature's "crisis disciplines": Does environmental communication have an ethical duty?', *Environmental Communication* 1(1) (May): 5–20.

Crompton, T. (2010) *Common Cause: The Case for Working with Our Cultural Values*, London: WWF-UK.

Cross, K., Gunster, S., Piotrowski, M. and Daub, S. (2015) *News Media and Climate Politics: Civic Engagement and Political Efficacy in a Climate of Reluctant Cynicism*, Vancouver, BC: Canadian Centre for Policy Alternatives.

Dietz, T., Fitzgerald, A. and Schwom, R. (2005) 'Environmental values', *Annual Review of Environment and Resources* 30: 335–72.

Dilling, L. and Moser, S. (2007) 'Introduction', in S. Moser and L. Dilling (eds), *Creating a Climate for Change: Communicating Climate Change and Facilitating Social Change*, New York: Cambridge University Press, pp. 1–27.

Douglas, M. and Wildavsky, A. (1982) *Risk and Culture: An Essay on the Selection of Technical and Environmental Dangers*, Berkeley: University of California Press.

ecoAmerica (2009) *Climate and Energy Truths: Our Common Future*, Washington, DC: ecoAmerica.

Entman, R. (1993) 'Framing: Toward clarification of a fractured paradigm', *Journal of Communication* 43(4): 51–8.

Feldman, L., Hart, P.S. and Milosevic, T. (2015) 'Polarizing news? Representations of threat and efficacy in leading US newspapers' coverage of climate change', *Public Understanding of Science*, DOI: 10.1177/0963662515595348.

Fischer, A., Peters, V., Vávra, J., Neebe, M. and Megyesi, B. (2011) 'Energy use, climate change and folk psychology: Does sustainability have a chance? Results from a qualitative study in five European countries', *Global Environmental Change* 21(3): 1025–34.

Griskevicius, V., Cialdini, R.B. and Goldstein, N.J. (2008) 'Social norms: An underemployed lever for managing climate change', *International Journal of Sustainability Communication* 3: 5–13.

Gunster, S. (2010) 'Self-interest, sacrifice and climate change: (Re-)framing the British Columbia carbon tax', in M. Maniates and J. Meyer (eds), *The Environmental Politics of Sacrifice*, Cambridge, MA: MIT Press, pp. 187–215.

—— (2011) 'Covering Copenhagen: Climate politics in B.C. media', *Canadian Journal of Communication* 36(3) (Fall): 477–502.

—— (2012) 'Visions of climate politics in alternative media', in A. Carvalho and T.R. Peterson (eds), *Climate Change Politics: Communication and Public Engagement*, Amherst, NY: Cambria Press, pp. 247–77.

Hackett, R.A., Wylie, S. and Gurleyen, P. (2013) 'Enabling environments: Reflections on journalism and climate justice', *Ethical Space: The International Journal of Communication Ethics* 10(2/3) (July): 33–46.

Hart, P.S. and Feldman, L. (2014) 'Threat without efficacy? Climate change on U.S. network news', *Science Communication* 36(3) (June): 325–51.

Hoggan, J. with Littlemore, R. (2009) *Climate Cover-Up: The Crusade to Deny Global Warming*, Vancouver: Greystone.

Howell, R.A. (2013) 'It's *not* (just) "the environment, stupid!" Values, motivations, and routes of engagement of people adopting lower-carbon lifestyles', *Global Environmental Change* 23: 281–90.

Hulme, M. (2009) *Why We Disagree about Climate Change: Understanding Controversy, Inaction and Opportunity*, New York: Cambridge University Press.

Immerwahr, J. (1999) *Waiting for a Signal: Public Attitudes toward Global Warming, the Environment and Geophysical Research*, American Geophysical Union. Accessed at http://research.policyarchive.org/5662.pdf.

Jackson, T. (2005) 'Motivating sustainable consumption', *Sustainable Development Research Network* 29 (2005).

Jamieson, K.H. and Capella, J. (2008) *Echo Chamber: Rush Limbaugh and the Conservative Media Establishment*, New York: Oxford.

Jones, M. (2013) 'Cultural characters and climate change: How heroes shape our perception of climate science', *Social Science Quarterly* 95(1) (March): 1–39.

Kahan, D., Jenkins-Smith, H. and Braman, D. (2011) 'Cultural cognition of scientific consensus', *Journal of Risk Research* 14(2): 147–74.

Klein, N. (2014) *This Changes Everything: Capitalism vs the Climate*, Toronto: Alfred A. Knopf Canada.

Kunda, Z. (1990) 'The case for motivated reasoning', *Psychological Bulletin* 108(3) (November): 480–98.

Lakoff, George (2010) 'Why it matters how we frame the environment', *Environmental Communication* 4(1) (March): 70–81.

Leiserowitz, A. (2005) 'American risk perceptions: Is climate change dangerous?', *Risk Analysis* 25(6): 1433–42.

—— (2006) 'Climate change risk perception and policy preferences: The role of affect, imagery, and values', *Climatic Change* 77: 45–72.

Leiserowitz, A., Maibach, E., Roser-Renouf, C., Feinberg, G. and Howe, P. (2013) *Global Warming's Six Americas, September 2012*, Yale University and George Mason University,

New Haven, CT: Yale Project on Climate Change Communication. Accessed at http://environment.yale.edu/climate-communication/article/Six-Americas-September-2012.

Leiserowitz, A., Maibach, E., Roser-Renouf, C., Feinberg, G. and Rosenthal, S. (2014a) *Public Support for Climate and Energy Policies in November 2013*, Yale University and George Mason University, New Haven, CT: Yale Project on Climate Change Communication.

—— (2014b) *Americans' Actions to Limit Global Warming, November 2013*, Yale University and George Mason University, New Haven, CT: Yale Project on Climate Change Communication. Accessed at http://environment.yale.edu/climate-communication/files/Behavior-November-2013.pdf.

Leiserowitz, A., Maibach, E., Roser-Renouf, C. and Smith, N. (2011) *Global Warming's Six Americas, May 2011*, Yale University and George Mason University, New Haven, CT: Yale Project on Climate Change Communication. Accessed at http://environment.yale.edu/climate-communication/article/SixAmericasMay2011.

Leiserowitz, A., Smith, N. and Marlon, J. (2010) *Americans' Knowledge of Climate Change*, Yale University, New Haven, CT: Yale Project on Climate Change Communication. Accessed at http://environment.yale.edu/climate/files/ClimateChangeKnowledge2010.pdf.

Lewis, J., Inthorn, S. and Wahl-Jorgensen, K. (2005) *Citizens or Consumers? What the Media Tell Us about Political Participation*, New York: Open University Press.

Lorenzoni, I., Nicholson-Cole, S. and Whitmarsh, L. (2007) 'Barriers perceived to engaging with climate change among the UK public and their policy implications', *Global Environmental Change* 17: 445–59.

Maniates, M. (2001) 'Individualization: Plant a tree, buy a bike, save the world?', *Global Environmental Politics* 1(3): 31–52.

Marshall, G. (2014) *Don't Even Think about It: Why Our Brains Are Wired to Ignore Climate Change*, New York: Bloomsbury.

McCright, A. and Dunlap, R. (2010) 'Anti-reflexivity: The American conservative movement's success in undermining climate science and policy', *Theory, Culture and Society* 27(2–3): 100–33.

—— (2011) 'The politicization of climate change and polarization in the American public's views of global warming', *The Sociological Quarterly* 52: 155–94.

Monbiot, G. (2015) 'We're not as selfish as we think we are: Here's the proof', *The Guardian* (October 14). Accessed at http://www.theguardian.com/commentisfree/2015/oct/14/selfish-proof-ego-humans-inherently-good.

Moser, S. (2010) 'Communicating climate change: History, challenges, process and future directions', *Wiley Interdisciplinary Reviews: Climate Change* 1 (January–February): 31–53.

—— (2011) 'Foreword', in L. Whitmarsh, S. O'Neill and I. Lorenzoni (eds), *Engaging the Public with Climate Change: Behaviour Change and Communications*, Washington, DC: Earthscan, pp. xv–xx.

Nerlich, B., Koteyko, N. and Brown, B. (2010) 'Theory and language of climate change communication', *Wiley Interdisciplinary Reviews: Climate Change* 1 (January–February): 97–110.

Nickerson, R. (1998) 'Confirmation bias: A ubiquitous phenomenon in many guises', *Review of General Psychology* 2(2) (June): 175–220.

Nisbet, M. (2009) 'Communicating climate change: Why frames matter for public engagement', *Environment: Science and Policy for Sustainable Development* 51(2) (March–April): 10–23.

Nisbet, M., Hixon, M., Moore, K. and Nelson, M. (2010) 'Four cultures: New synergies for engaging society on climate change', *Frontiers in Ecology and the Environment* 8: 329–31.

Nisbet, M. and Myers, T. (2007) 'Twenty years of public opinion about global warming', *Public Opinion Quarterly* 71(3) (Fall): 444–70.

Norgaard, K.M. (2011) *Living in Denial: Climate Change, Emotions, and Everyday Life*, Cambridge, MA: MIT Press.

Ockwell, D., Whitmarsh, L. and O'Neill, S. (2009) 'Reorienting climate change communication for effective mitigation', *Science Communication* 30(3) (March): 305–27.

Olausson, U. (2009) 'Global warming–global responsibility? Media frames of collective action and scientific certainty', *Public Understanding of Science* 18(4): 421–36.

O'Neill, S., Williams, H., Kurz, T., Wiersma, B. and Boykoff, M. (2015) 'Dominant frames in legacy and social media coverage of the IPCC Fifth Assessment Report', *Nature Climate Change* 5 (April): 380–5.

Oreskes, N. and Conway, E. (2010) *Merchants of Doubt: How a Handful of Scientists Obscured the Truth on Issues from Tobacco Smoke to Global Warming*, New York: Bloomsbury.

Owens, S. (2000) '"Engaging the public": Information and deliberation in environmental policy', *Environment and Planning A* 32: 1141–8.

Peattie, K. (2010) 'Green consumption: Behavior and norms', *Annual Review of Environment and Resources* 35: 195–228.

Potter, E. and Oster, C. (2008) 'Communicating climate change: Public responsiveness and matters of concern', *Media International Australia: Incorporating Culture and Policy* 127: 116–26.

Pratt, S. (2015) 'Action on climate change backed: Poll', *Edmonton Journal* (September 30): A4.

Roeser, S. (2012) 'Risk communication, public engagement, and climate change: A role for emotions', *Risk Analysis* 32(6): 1033–40.

Rootes, C. (2007) 'Acting locally: The character, contexts and significance of local environmental mobilisations', *Environmental Politics* 16(5): 722–41.

Roser-Renouf, C., Stenhouse, N., Rolfe-Redding, J., Maibach, E. and Leiserowitz, A. (2015) 'Engaging diverse audiences with climate change: Message strategies for global warming's six Americas', in A. Hansen and R. Cox (eds), *Routledge Handbook of Environment and Communication*, New York: Routledge, pp. 368–86.

Schmidt, A. and Schäfer, M. (2015) 'Constructions of climate justice in German, Indian and US media', *Climatic Change*, 133(3): 535–49.

Senecah, S. (2007) 'Impetus, mission, and future of the Environmental Communication Commission/Division: Are we still on track? Were we ever?', *Environmental Communication* 1(1): 21–33.

Shaw, A., Sheppard, S., Burch, S., Flanders, D., Wiek, A., Carmichael, J., Robinson, J. and Cohen, S. (2009) 'Making local futures tangible–synthesizing, downscaling and visualizing climate change scenarios for participatory capacity building', *Global Environmental Change* 19: 447–63.

Shove, E. (2010) 'Beyond the ABC: Climate change policy and theories of social change', *Environment and Planning A* 42: 1273–85.

Shue, H. (2015) *Climate Justice: Vulnerability and Protection*, New York: Oxford University Press.

Steg, L. and Vlek, C. (2009) 'Encouraging pro-environmental behaviour: An integrative review and research agenda', *Journal of Environmental Psychology* 29(3) (September): 309–17.

Stern, P. (2000) 'Toward a coherent theory of environmentally significant behavior', *Journal of Social Issues* 56(3): 407–24.

Stoknes, P.E. (2015) *What We Think about When We Try Not to Think about Global Warming: Toward a New Psychology of Communication*, White River Junction, VT: Chelsea Green Publishing.

Taylor, D. (2000) 'The rise of the environmental justice paradigm: Injustice framing and the social construction of environmental discourses', *American Behavioral Scientist* 43(4) (January): 508–80.

The Topos Partnership with Pike, C. and Herr, M. (2009) *Climate Crossroads: A Research-Based Framing Guide*, The Topos Partnership.

Webb, J. (2012) 'Climate change and society: The chimera of behaviour change technologies', *Sociology* 46(1): 109–25.

Whitmarsh, L., O'Neill, S. and Lorenzoni, I. (2013) 'Public engagement with climate change: What do we know and where do we go from here?', *International Journal of Media & Cultural Politics* 9(1): 7–25.

Wood, B.D. and Vedlitz, A. (2007) 'Issue definition, information processing, and the politics of global warming', *American Journal of Political Science* 51(3) (July): 552–68.

Chapter 3

Environmental protest, politics and media interactions

An overview

Susan Forde

Introduction: The politics of environmental protest

When we consider the relationship between environmental protest and the media – both in the modern digital era and in times past – our first consideration must be for the notion of power. As with most interrogations of the place of media in democracy, 'power' and its place of residence is fundamental to understanding the interactions between the key players – politics, the public and the media. In Chapter One, we examined the different models of democracy that might exist and the conditions they create to facilitate and possibly enhance civic activity and protest. Two key Australian authors in this field of environmental protest and the media, Brett Hutchins and Libby Lester (2006), note the relationship between environmental protest and the media is 'never a relationship of equal power', with environmentalists regularly trying to 'keep up' with media agendas and angles (p. 434). They draw heavily upon Manuel Castells' theorizations of power and dominance in the context of the information society, in which he characterizes environmentalism as 'a pervasive social movement . . . [that] constitutes a principal site of resistance to global capital and the domination of social life by economic interests, or the new power system' (Castells 2004: 182; Hutchins and Lester 2006: 435).

If we are to accept this characterization of the environmental movement, then it has much in common with many of the radical social movements that have emerged particularly in Western democracies in the 20th and 21st centuries – Indigenous land rights, the working-class/socialist movement, the Occupy campaigns and so on. It is important, however, that for us properly to understand environmentalism as a social movement, and thereby appropriately to analyze its relationship with media, we must also accept the diversity that exists in environmental protest – while some at its core challenge powerful institutions and capital, other protests feature a not-insignificant number of self-interested protesters (Machin 2013) who feature in particular campaigns. Indeed, Chapman *et al.* note that the environmental movement is a progressive movement that is, strangely, not considered to be party-political, gaining somewhat similar support from both leftist and conservative voters alike, depending on the issue (1997: 38). So, while it similarly carries the challenges to landed power that other radical movements carry, the environmental movement – perhaps like some (but not all) social movements – has an added dimension of featuring a range of motivations from the protesters involved. This chapter, then, considers environmental protest and opposition in its diversity and, in doing so, considers the different roles that the media may fulfill in covering environmental conflict and in representing climate crisis.

Gamson and Wolfsfeld's 1993 study of the power relationship between social movements and media provides insight into the different angles that the environmental movement takes in diverse circumstances and helps explain the sometimes apolitical nature of environmental politics. Their words help us understand the fine line that the movement walks, sometimes supporting a somewhat safe political option; other times, directly challenging powerful interests and structures:

> The environmental movement provides examples of the risks involved in seeking broad public support. Some groups avoid targeting corporate or governmental actors, focusing their attention on consumer behavior instead. Recycling or anti-littering campaigns, for example, make few enemies and put the movement on the side of the angels. But the favorable image is purchased at the cost of leaving broader structural and cultural sources of environmental problems unchallenged and implicitly reinforced.
>
> (1993: 123)

An editorial collective member from a radical environmental publication in the U.K., *Do or Die*, distinguished between two types of environmental activists (Atton 2002): the professional activists who work for organizations such as Friends of the Earth and Greenpeace 'and who take part in large-scale actions to raise awareness of issues through media exposure' and 'ordinary people' who undertake direct action to affect change. Amanda Machin's important work on the centrality of radical democracy to climate change 'solutions' (and she emphasizes that there is no one solution, nor the possibility of consensus) notes the 'techno-economic' initiatives related to climate change which force short-term action but which may, in the end, all be carried out for 'self-interested' reasons rather than sustained commitment to the broader goal. While policies may develop new environmentally friendly technologies, encourage recycling, offer green taxes and coerce industry with emissions allowances, these policies cannot 'by themselves, produce any real structural change':

> They are often invisible and demand no awareness or engagement of the dangers of climate change by the population (take away the tax or the emissions cap, it seems, and behaviour will revert). These sorts of policies can be criticized for actually affirming the very self-interested behaviour that is at issue.
>
> (Machin 2013: 3)

This is a core challenge for the environmental movement, and in this way environmental activism – or perhaps, environmental action – can be set apart from some other radical social movements that may not offer the same opportunities for self-interest. Environmental activism should be undertaken not by people who self-perceive as individuals, but as citizens, enacting behaviour for (perhaps) a broader political goal rather than for what it will deliver to their own interests. This stance is represented by Machin's 'green republicans' but they, too, rely on a consensus model, assuming that there will be a united approach to combatting climate change and that the public will eventually subscribe to this 'solution' (Machin 2013). Kenis and Lievens summarize this positioning well, noting the more recent trend to 'depoliticize' environmental politics primarily represents an attempt to find 'greener versions of the modern market, and growth-oriented, liberal democratic model' (2014: 532; the authors also draw on Blühdorn 2007, 2013).

There is evidence, however, that some environmental protest, at its core, challenges powerful institutions and is subsequently relegated to the same standing as, for example, anti-capitalist protest, Indigenous land rights actions, feminist protest and so on. Hutchins and Lester's examination of the pioneering Franklin Dam blockade in Australia in the early 1980s demonstrated where the line is drawn:

> Different rules of engagement, and types of critique, are applied to environmentalists and political activists promoting humanist values and the sanctity of place when compared to celebrities, politicians and prominent business-people (usually men) promoting themselves, their power and/or the interests of profit and capital. The common denominator between these groups is the attempted control of a media message, but it is their relationship with media organizations and journalists that differs. Those advocating or acceding to the interests of capital are more likely to be located within the privileged formation of the space of flows.
>
> (2006: 446)

Chapman *et al.* identify that the political agenda is a key driver in the coverage of the environment, and is a 'major reason' for the appearance and disappearance of environmental issues from the media (and public) agenda (1997: 43). They cite the case of former British Conservative Prime Minister Margaret Thatcher, who highlighted a range of environmental issues during her term in an attempt to 'claim back' the environmental space for conservative politics. The initial impetus for this argument may well have been Thatcher's determined attempts to attack the legitimacy of striking coal miners. But the environment correspondent at Britain's Channel 4 news found the Conservatives' agenda soon changed – and the media followed – when the real impact of successful environmental campaigning became clear:

> There was the realisation in Conservative-supporting newspapers that you could report the environment massively as long as it was perceived in the

ranks of the Conservative Party that the environment was important. . . .
When green issues came to be seen as a threat to financial interests, then the
enthusiasm declined suddenly. In other words, as soon as the environment
correspondents dug in and started looking for the real causes of things, then
Conservative editors became rather less keen on running these stories.

> (1997: 44, citing Channel 4 environment reporter Andrew Veitch)

This chapter is designed to provide an overview of the interactions between the
environmental protest movement and the media. It selectively draws from a body
of previous literature which provides case studies, theoretical positioning and
more systematic empirical studies of the methods utilized by the environmental movement to raise awareness for its issues through media relationships and
subsequent coverage. The literature encompassing environmental protest and the
media is vast, and could not possibly be canvassed fully in this chapter. To that
end, then, I summarize some historical features of the environmental movement's
interactions with the media, to suggest what this tells us about the *media's role* in
presenting the environmental movement to the public, and to therefore point to
successes and failures which might guide future strategies. It concludes by summarizing the relationship among environmental protest, the media and politics,
paving the way for subsequent chapters to present new case studies and theorizations around journalistic paradigms for climate crisis reporting.

Environmental protest, crisis and the media

In broadly considering the interactions between social movements and the media,
Gamson and Wolfsfeld identify a 'drearily predictable' interaction:

> 'Send my message,' say the activists; 'Make me news,' say the journalists.
> In this dialogue of the deaf, neither activists nor journalists make an effort to
> understand how the other views their relationship or, better yet, the complex
> nature of these transactions.
>
> (1993: 115)

Tellingly, Gamson and Wolfsfeld's structural and cultural analysis of the social
movement/media relationship finds the movement far more dependent on the
media than vice versa, meaning that 'this fundamental asymmetry implies the
greater power of the media system in the transaction' (1993: 115). While public
discourse about the movement and its activism is carried out in various forums,
including the movement-related media – alternative media – most of the people
the movement wishes to reach are part of mass media audiences who are 'missed'
by the movement's own meetings and media outlets. Reaching a broader media,
then, becomes integral and essential to the movement's clear goals to influence
a broader audience beyond their own advocates. Gamson and Wolfsfeld suggest
that broader media coverage 'validates' the movement – '[t]he media spotlight

validates the fact that the movement is an important player' (1993: 115). An activist event that attracts no media coverage, they argue, is a 'nonevent' because it has little scope to influence the public or its intended target (policy-makers, decision-makers, developers): 'no news is bad news' (1993: 115). Luis Hestres offers an important intervention here. His work on Internet-based environmental activist groups such as 350.org and 1Sky notes the tendency for these organizations to make deliberate attempts to 'preach to the choir' – to target, for mobilization, those publics that they already know to be engaged with the issue and indeed 'alarmed' about climate change (2014: 330). This approach challenges previous strategies – outlined by much of the literature in this chapter – to use 'mainstream' legacy media to reach a broad-based public, and represents an important emerging field of research for climate change social movements. This strategy is considered in Chapter Five.

Hutchins and Lester's (2006) work follows and builds upon much of the environmental communication scholarship, drawing upon Castells (2004) in enunciating the important relationship between media and public perceptions of the environment, identifying news outlets as 'the avenue through which the majority of the public becomes aware of impending environmental threats and economic developments' (2006: 434). As with most protest movements, environmentalists see the importance of the media in conveying their message to a broader public, and journalists pick up the stories as they recognize they will be of interest to a readership looking to protect their local environment and well-being (2006: 434). Simon Cottle refers to 'global crises', a range of internationally felt and reported crises that are well-known to most – the global financial crisis, SARS, terrorism, the world food crisis caused by debilitating war and mass displacement, and 'environmental despoliation and threats to biodiversity and, of course, climate change' (2011: 78). He distinguishes a 'global' crisis from (perhaps) an 'ordinary' crisis as one:

> whose origins and outcomes cannot for the most part be confined inside the borders of particular nation states; rather, they are *endemic* to, *enmeshed* within, and potentially *encompassing* of today's late-modern, capitalistic world – a de-territorializing world that has become increasingly interconnected, interdependent and in flux, that is to say, *globalized.*
>
> (2011: 78; emphasis in original)

He considers that media and communications scholars must, in a more theoretical and empirical way, begin to offer understandings of the place of news and journalism in existing and evolving crises so that these media may become more a part of the creation of a recognized sense of 'globality' – enmeshing their audiences in a process of acceptance which offers future solutions (2011: 78ff). Importantly, global crises not only extend beyond national borders, but imply an 'interpenetrating complexity and truly catastrophic nature' which gives rise to a sense of 'impending apocalyptic collapse' (p. 79). This in turn, Cottle suggests, evokes a sense that we are perhaps dealing with not an end to the world, but an end to the world as we know it. Cottle (2011; and also Lester and Cottle 2009) examined the coverage of the 2007 release of the Intergovernmental Panel on Climate Change report which formally recognized the reality of global warming. He found that the news media, for the first time, enunciated a clear and compelling discourse around climate change, bringing it into the homes of a global audience despite the best efforts of the powerful climate change sceptics lobby:

> In such culturally resonate [sic] ways, the impinging realities of climate change, hitherto largely invisible or contained in the expert discourses of science and politics, became symbolized, visualized and dramatized and, for a political moment at least, turned into a high-profile, 'knowable' and, potentially, actionable global crisis.
>
> (Cottle 2011: 90)

Hutchins and Lester confirm the contribution that 'media flows' make to grassroots politics such as that found in environmental organizations. These organic protest groups 'create and organize themselves in network formations' that assist them to disseminate information, raise awareness of issues and coordinate action. 'In other words, to generate knowledge of grassroots politics, a concerted effort must be made to engage with and move within the space of flows and, more specifically, the space of media flows' (2006: 437). Importantly, Hutchins and Lester address the instinct within the environmental movement (and indeed, most political social movements) to stay within their own comfort zone – to refuse to engage with dominant media and their associated interests and to reinforce communications within the movement. This would classically be considered 'preaching to the converted'. The environmental movement has, since the 1960s, successfully infiltrated mainstream media agendas in successful ways, and this is due to a recognition that they have 'little choice' but to engage:

> The news media is the key structuring intermediary in the conduct of public affairs, and political communication and information are transmitted in the space of media flows. The choice to stay outside this space may offer the comfort of ideological purity, but the companion of this condition is marginality. Without the widespread awareness generated by news coverage, environmental action and values lose both legitimacy and effect, failing to appear on mainstream political and cultural agendas and register in the collective mind.
>
> (2006: 437; and citing Castells 2000: 365)

They recognize the importance of positive coverage of protest, even if that protest is a short-lived 'burst' of activity, as news coverage that meets the agenda of the environmental movement 'has lasting impact on the collective perception of an issue' (2006: 438). While Hansen (1993) is concerned with the longer-term strategies of the environmental movement to set agendas and create influence,

Hutchins and Lester are identifying the value in positive coverage of stand-alone events. Hansen (2010) notes our understanding of an 'environmental discourse' – public discussion about environmental issues – is a relatively recent phenomenon, and distinct from earlier historical discussions about notions of 'conservation'. Indeed, Carson's work *Silent Spring* from 1962 is identified as the earliest foray into enunciating what we know as a discourse about environmentalism, hence both the intertwined media and public discussions of the environment hark back only about 50 years or so, to the 1960s (2010: 1). Solesbury notes in 1976 that while attention to issues around pollution and protection of land had been evident for some 150 years, it was only in recent times that 'a new class of issue has become recognized – environmental issues' (1976: 380). DeLuca (2011) recounts one of the early successful actions of Greenpeace, placing themselves between a whale and a Soviet whaling ship in an effort to protect the whale in June 1975. The six activists in an inflatable Zodiac with a film camera immediately released the footage to media and, while the whale was not saved, Greenpeace had 'entered the mass consciousness of modern America' (2011: 177, citing Greenpeace activist Robert Hunter). DeLuca identifies that with this anti-whaling action, Greenpeace introduced 'direct action image events into the repertoire of environmental activism':

> Instead of sublime landscape photographs designed to inspire letter writing and lobbying, videotaped image events capture people intervening on behalf of nature. This move from scene to action is facilitated by advances in video camera technology and marks an epistemic break in ways of knowing the world.
>
> (2011: 177)

The Greenpeace action was followed by subsequent tree-sitting events to protect ancient forests, attracting major media attention because of the novelty of the event (which has since been repeated by environmentalists around the world). The longest tree-sit event featured ex-waitress Julia 'Butterfly' Hill, who lived in a 1,000-year-old redwood tree for two years from 1997 to 1999 'to successfully articulate the inextricable link of wilderness and social issues' (2011: 177). Hutchins and Lester use as their case study the coverage of the Franklin Dam blockade in the Tasmanian wilderness, which they identify as 'the first wilderness campaign to obtain global stature' (2006: 439; and citing Hay 1991–92). The Wilderness Society, now a stalwart of environmental action in Australia, was at the time a young organization, which was successful in tapping into global ideas about the environmental movement and:

> resistance to capital. . . . At the end of this chain were the news media and its workers'. . . [who were] charged with reporting the intricate dynamics between the rights of capital, the emerging politics of environmentalists, and the spectacle and tactics of protests. We argue that this event, occurring at a

formative point in the information age, established a pattern of environmentalist-media interaction that was to have lasting influence in the Australian context.

(Hutchins and Lester 2006: 440)

They identify the Franklin Dam blockade as a key moment which illuminated the tenuous relationship between protesters and the media, with each struggling to control the news agenda. Protesters used theatrical stunts and photograph-worthy actions to elicit front-page coverage. They were, as with the original Greenpeace 'anti-whaling' protest mentioned previously, attempting to enter the Australian consciousness with their high-profile action. The ongoing arrest of protesters – in total 1,227 arrests and 447 imprisonments by the time the blockade ended – appealed to basic news principles of conflict and timeliness, ensuring ongoing dramatic coverage of the confrontations among the hydroelectric company attempting to dam the Franklin River to generate electricity, the police and the protesters.

Gamson and Wolfsfeld (1993) confirm that social movements provide media with conflict, drama, action, colourful copy and photo opportunities. Over time, however, journalists covering the Franklin Dam blockade tired of the environmental 'stunts'; coverage halved, and by the end of the three-month-long blockade, journalists had become sceptical about the environmentalists' aims and began to see each new day of action as an attempt to garner media coverage rather than an overwhelming commitment to the cause of saving the river (Hutchins and Lester 2006: 444). Following a number of other protest actions to protect old-growth forests in Tasmania, the Wilderness Society pulled back from its tactics to attract attention through newsworthy 'stunts':

> Direct action, conceived of as a short, sharp attack on the news agenda, was no longer considered an effective protest tool, and the Wilderness Society announced an end to these tactics.

(p. 445)

Chapman *et al.* found that most environment reporters were initially 'sympathetic' to the cause, creating a level of monitoring among the editorial decision-makers to ensure their copy was more rigorously considered so that the news organization was not seen to have a soft line on environmental issues (1997: 38–9). This is consistent with Hutchins and Lester's (2006) findings that, once the media had a sense they had been 'played', to some extent, by the environmental movement, they became increasingly sceptical and increasingly oppositional to future environmental protest. Indeed, the collusion that many journalists felt in the early days of the famous Franklin Dam blockade seemed to cause a later opposite reaction, in that they became *less likely* to run the message of environmentalists in subsequent campaigns and mocked the staged events, likening them to a Hollywood film set. Chapman *et al.* similarly recognize that environmental groups are 'skilled in tactics which provide good visual news material' and note that environmental campaigner Greenpeace 'has often led the way by taking dramatic high-seas action in little speed-boats to spoil Japanese whaling expeditions – or protesters in the path of motorways hang themselves from the tree tops or from high tetrapods' (1997: 119).

Essentially though, we might see that while such tactics attain immediate media attention, they do not necessarily point to ways in which the environmental movement might control the news agenda or determine how their cause is framed and interpreted. This is a core concern of much of the literature around environmental protest strategies and the media. Simply, in Solesbury (1976), there is a marked difference between the media tactics of social movements, such as environmentalism, which might 1) command attention, 2) claim legitimacy and 3) invoke action. He sees these three phases of the environmental movement (which may or may not all occur in a particular campaign) as 'three tests which the nascent issue must pass in order to remain on the agenda for debate and decision' (p. 395).

There was a real lull in media interest in environment campaigns throughout the late 1980s and 1990s, and indeed Chapman *et al.* cite *The Guardian*'s Paul Long who claimed news editors 'around the world yawned when the [1995] World Climate Conference in Berlin drew to a close' (in Chapman *et al.* 1997: 36). This was despite the fact that the famous 'Brent Spar' case had occurred in 1995, representing a strong campaign by Greenpeace against oil giant Shell, who wanted to dump the redundant Brent Spar oil-storage installation into the Atlantic Ocean. Consistent with the general malaise in environmental reporting at the time, this significant environmental campaign generally received negative coverage for Greenpeace, with Greenpeace's politics labelled 'environmental jihad' by the British press (Hansen 2010: 42). Anderson and Gabor similarly detected a lack of interest, particularly among television journalists, in the major Rio Earth Summit of 1992, identifying 'that delightful American-invented condition "antisappointment" – too much hype, too little substance, which in turn led to a boredom factor, or "environmental overfeed" as one editor described it' (1993: 50). Notably, journalists' attitudes were out of step with the public's, according to Anderson and Gabor, as the public appetite for environmental stories remained strong. A BBC television news journalist reported that environmental activists were 'very well aware of how to manipulate the media' – often through providing cost-effective video packaging (now standard practice, but a newer concept at the time of the study in the mid-1990s) which journalists felt was an attempt to manipulate their agendas while at the same time destroying environmental lobbyists' credibility among journalists (1993: 50). This historical context is pointing to two things – first, that environmental activists have a track record in creating newsworthy, attention-grabbing stunts which make the news. In doing this, they are attempting to 'enter the consciousness' of the public with their issue. Second, however, this propensity to attract media coverage does not necessarily translate long-term into a successful campaign, nor does it create positive relationships between activists and journalists, as the latter often feel manipulated by the regular 'events'.

Recent work completed at the Reuters Institute for Journalism at the University of Oxford (Painter 2011) found that from 2007 to 2010, there was a noticeable rise in the inclusion of 'sceptical' voices in dominant media reportage of climate change. This major cross-national study found, importantly, that sceptical voices were far more likely to appear in 'Anglo-Saxon' countries such as the United States, the United Kingdom and Australia compared with other nations such as Brazil, China, France and India – indeed, the 'Anglophone' countries accounted for more than 80 percent of sceptical voices about climate change (2011: 2). This seemed to be because these countries were far more likely to present the views of politicians, representing ideological stances on climate change and global warming, and secondly, these voices were far more likely to appear in right-leaning rather than left-leaning media outlets (pp. 2–3). These countries were also more likely to host climate change sceptic organizations – such as the U.K.'s curiously named Global Warming Policy Foundation – and journalistic routines called for these voices to be covered. Painter concludes:

> The weight of this study would suggest that, out of this wide range of factors, the presence of politicians espousing some variation of climate scepticism, the existence of organised interests that feed sceptical coverage, and partisan media receptive to this message, all play a particularly significant role in explaining the greater prevalence of sceptical voices in the print media of the USA and the UK.
>
> (p. 5)

While Painter's findings about the dominant media space given over to climate change sceptics refer primarily to the U.S. and the U.K., he found strong trends in Australia and Canada. In the former, this was due primarily to the right-leaning, conservative newspapers run by Rupert Murdoch's News Ltd.; in Canada, prominent space was given to climate sceptics hailing from Alberta, the site of the country's major tar sands deposits (Painter 2011: 40–1). In countries such as China, India, Peru and on the African continent – indeed, in most developing countries – there was an overwhelming acceptance that climate change was attributable to anthropogenic causes and there was very little or no coverage of the sceptics' arguments (pp. 42–3).

Painter's study, and much of the literature considered so far, relies on the coverage contained in the so-called 'legacy' media – traditional forms of news such as television, radio and newspapers. Many of the techniques used in this era of traditional media have carried through to the digital age. Anti-globalization protests featuring environmentalists, labour and economic justice activists (Juris 2005: 193) found high-energy visual images and media 'events' to be tried and true ways to obtain media saturation for their politics, which gained significant momentum after the 1999 World Trade Organization protests in Seattle. Juris notes such anti-corporate globalization movements feature three broad characteristics – while movement networks may be locally rooted, the issue and activism is global in nature; they are organized around multiple virtual and physical network forms; and they are *informational*:

> The various protest tactics employed by activists, despite emerging in different cultural contexts, all produce highly visible, theatrical images for mass mediated consumption.
>
> (2005: 195)

Juris finds contemporary grassroots activists, such as those involved in the WTO and Occupy movement actions, have developed 'highly advanced forms of computer-mediated alternative and tactical media' (p. 204), including Internet open-source news site Indymedia, digital forms of culture jamming, hacktivism and electronic civil disobedience (e.g. flooding particular websites in order to overwhelm their servers and subsequently shut down the sites). Importantly, Juris argues these digital activist actions 'have facilitated the emergence of globally coordinated transnational counterpublics while providing creative mechanisms for flexibly intervening within dominant communication circuits' (2005: 204; see also Hestres 2014). This could be partially because activist environmental organizations and their media were early adapters of digital technology and publishing. Chris Atton identified PeaceNet in the United States and Green Net in the U.K. as two of the earliest Internet-based media projects (2007: 59), while the EnviroWeb delivered some of the earliest forms of downloadable alternative media. Hutchins and Lester confirm the activity of environmental activists – as with those working in all social movements – has adjusted noticeably with the advent of the Internet, and particularly social media. They tell of the 'increasingly complex structure' of activism, with high-profile staged events now coordinated through online networks and connected action (2015: 338–9).

Media 'roles' in communicating the environment

We can draw upon Christians *et al.* (2009) to understand media behaviour in covering environmental issues and conflict. While we can fairly simply describe the role of alternative and environmental media outlets covering these issues as falling within the 'radical' role of the media, the function of broader, dominant media rejects such overt mobilization of audiences that the radical role requires. As discussed in our introductory chapter, Christians *et al.* note four substantial traditions in the history of debate on the norms of public communication – corporatist, libertarian, social responsibility and citizen participatory. Each tradition 'implies a set of institutions that time has tested, and that continue to be an important source of norms for the roles of journalism in democracy' (2009: 21). Couldry and Curran note that as media power is recognized as an 'increasingly significant theme of social conflict', the focus of study must shift beyond mainstream productions such as major television, radio, online and print forms to the 'wider terrain of media production, some of which seeks, explicitly or implicitly, to challenge central concentrations of media resources' (2003: 7). In this vein, Christians *et al.*'s citizen participatory *tradition* of public communication emphasizes the role of the local community, along with small-scale and alternative media. They note:

> In fact, there is an intrinsic difficulty in applying this tradition's thinking to extensive, mainstream national or international media like network television or the mass press. Nonetheless, this perspective furnishes a critique of such media and sets up certain criteria of desirable operation. Even large-scale media can have a concerned and responsive attitude to their audiences and encourage feedback and interactivity. They can employ participatory formats and engage in surveys and debates that are genuinely intended to involve citizens.
>
> (Christians *et al.* 2009: 25)

Nonetheless, the citizen participatory tradition does not aptly describe the behaviour of the media in relation to environmental conflict, even though it is appropriate to our understanding of the ways alternative, independent and radical media might have negotiated public communication around environmental debates. Similarly, in identifying the four consistent roles of 'media in society' over time, Christians *et al.* note the *radical* role as one of the four which again describes alternative and independent (most often supporting the environmental movement) but does not adequately explain the activities of the dominant media covering, for example, the Franklin Dam blockade, the tar sands in Canada or the very current debates around destruction of coral in the Great Barrier Reef. Rather, the 'monitorial' and 'facilitative' roles more accurately describe this media activity. The *monitorial* role is based on Lasswell's notion of 'surveillance' – a 'scanning' of the world for relevant information, events, conditions to be imparted (2009: 140; and see Lasswell 1948). It is a classic 'objective' or 'neutral' role for journalism – reporting key events factually; providing a guide to public opinion and the views of key players and opinion leaders; and covering key institutions such as Parliament, the courts, press conferences of significant bodies and so on (Christians *et al.* 2009: 145). The limitations of this description of media activity are not the concern of this chapter (and Christians *et al.* recognize its faults); rather, we argue that in attempting to cover the environment in a 'balanced' or 'neutral' way, many dominant media journalists could consider that they are fulfilling a 'monitorial' role. This plays out, for example, in the aforementioned equal legitimacy given to climate sceptics and climate change scientists (Painter 2011).

The *facilitative* role is evident in some sections of the media in covering the climate change crisis, and this is confirmed by past research consulted here. In this role, the media are more active than in their monitorial state, reflecting the political order and focusing on civic democracy (Christians *et al.* 2009: 158ff). The media promote dialogue among their readers and viewers 'through communication that engages them in which they actively participate. In facilitative terms, the news media support and strengthen participation in civil society outside the state and the market' (p. 158). They identify environmental politics as a clear example of a 'social conflict' that requires the media to carry out a facilitative role:

> Social conflicts are a major component of democratic life, and in delibera-
> tive politics they remain the province of citizens rather than of judicial or
> legislative experts. Affirmative action, environmental protection, health
> care policy . . . global warming, gun control, arms trade, welfare reform and
> doctor-assisted suicide . . . raise moral conflicts that the public itself must
> negotiate.
>
> (p. 159)

The facilitative role is recognized by the historical and liberal Hutchins Commission (United States), enunciating a socially responsible press which provides
truthful and intelligent information to facilitate public engagement rather than
the direct 'recount' of a monitorial press (Commission on Freedom of the Press
1947). This form of news is defined by its obligations to the community and its
remit to create a healthy society rather than to focus on the rights of individuals to
publish (Christians *et al.* 2009: 160). This facilitative role has much in common
with the notion of Public Journalism, also termed Civic Journalism or Community
Journalism, following a tradition of social responsibility theory and development
journalism (p. 161). It is a *reforming* role for the media rather than a radical role.
There is evidence from the literature that dominant media reportage of environmental conflict (e.g. early coverage of the Franklin Dam blockade as outlined by
Hutchins and Lester 2006) sometimes reflects this facilitative role in that it gives
prominence to the actions of environmental activists, and this implies the facilitation of action by others. The studies presented in this chapter would suggest that,
in the cases when the media do carry out a facilitative role in relation to environmental politics and conflict, this role is often short-lived and quickly reverts
to either a straight monitorial role or an open aversion to the messages of the
environmental movement. Journalists become sceptical of the ongoing 'stunts'
designed to attract headlines, with a new event each day of the campaign.

Consistent with Christians *et al.*'s facilitative role, Cottle (2006) highlights
an assumed 'constitutive role' for the media in the experience of conflicts and
the subsequent political fall-out (also in Hutchins and Lester 2015). The media have
a role well beyond representation and reporting and take an active role in the
'making' and heightening of conflict (Cottle 2006). Hutchins and Lester's more
recent work (2015) refines this concept further, identifying 'mediatized environmental conflict' which they argue is a necessary and specific conceptual advance
because of the 'political significance of the environment, and the pivotal role of
media in contests over the definition and understanding of environmental risks
and impacts' (2015: 341). They offer an important contribution in identifying
a 'four-phase' approach to mediatized environmental conflict. The latter term
refers to the fact that in a media-saturated world, environmental conflict is constantly communicated in relation to environmental risks, threats and disasters

(p. 339; and drawing on Beck 2009; Cottle 2006; Pantti *et al.* 2012). Hutchins and Lester's four-phase approach again uses Tasmania as its case study and theorizes mediatized environmental conflict to be 'constituted by the interactions occurring between four key spheres of action: (i) activist strategies and campaigns, (ii) journalism practices and news reporting, (iii) formal politics and decision-making processes, and (iv) industry activities and trade' (2015: 339). Mediatized conflict is, therefore, 'a product of the mutually constitutive interactions between activism, journalism, formal politics and industry' (2015: 339). In essence, the activities of all four spheres of activity around mediatized environmental conflict are conscious of and pay heed to the anticipated response of the other spheres (p. 342).

A radical media role for climate crisis?

In establishing normative roles for the media, Christians *et al.* outline their notion of a 'radical' function for media in society. To revisit this briefly from our Introduction, the radical role is recognized as existing primarily in competitive market societies where the free market system usually leads to significant inequalities in power, opportunity, education, social status and wealth. While the discourse of the 'system' suggests that these inequalities develop as a result of personal initiative (or lack thereof), Christians *et al.* identify that journalism carrying out this role 'ensures that no injustice is ever tolerated' (2009: 179). The radical role assumes that 'power holders impede the flow of information' and that the system of public communication needs reform if less powerful groups are to be represented and adequately informed.

Importantly to this chapter, the role of the radical journalist goes further than this to propose a new order and to '*support movements opposing these injustices*' (2009: 179; emphasis added). The radical role recognizes that some alternative and community media that form part of this sector are less radical and more a part of the broader socio-political system; these might include ethnic community radio, for example, or community-based media that are less politically radical but still designed to give voice and empower minority groups. There is some crossover between the 'facilitative' and 'radical' role of journalism in this space, but the literature presented here suggests the radical role is well beyond the reach of dominant media and even some community media organizations – and that it can only be carried out in radical publications supporting the marginalized, often minority, interests. Such publications overtly '[side] with those who are developing forms of resistance and advocacy against the dominant power holders' and are 'by definition partisan' (p. 180). Furthermore, where dominant media journalists might attempt to 'obscure' the conflicts of interest between those who dominate political-economic conditions and those who have no influence over these conditions, the radical journalist will expose this conflict of interest and point out the injustices and contradictions in it (pp. 179–80).

Conclusion

It is an interesting note that in 1997, at the time Chapman *et al.* produced a cross-national study on environment and the media in Britain and India, coverage of the environment was considered a 'luxury' – an issue that was covered in times of prosperity but that quickly disappeared from the political and media agenda in times of economic difficulty. One of the reporters interviewed for Chapman *et al.*'s study clarified that 'they [the people] like to talk about recycling and doing green things when there is no threat of the old man losing his job. But I am afraid that when the economy is in a muddle it just goes down the pecking order of events' (1997: 45). The current climate crisis – sometimes referred to as a planetary emergency – has now overtaken that previous approach, with coverage of the environment and recognition of the urgency of the issue persisting, to at least some extent, regardless of the prevailing economic circumstances.

Foxwell-Norton (2015) identifies that the greatest opportunity for sound coverage of environmental issues rests with community and alternative media forms – indeed, the proximity of community media producers to their local communities ensures a safer pair of hands, so to speak, than that offered by often-distant and preoccupied dominant media. She defines community media in a geographical sense, relating to place:

> [that is], community media that have a local physical presence in communities and are best positioned to communicate community participation in local environmental management and issues. . . . It is their capacity and intent (however successful) to represent and reproduce the everyday lived experiences of community at the local level . . . which offers a plethora of opportunities for the communication of local and global environmental issues.
>
> (p. 391)

Hutchins and Lester wrote of the 'ceaseless dance between activists, journalists, politicians, and industry leaders' occurring in the ongoing fights for the Tasmanian forests and wilderness – battles that have continued long after the original Franklin Dam blockade which first brought the international gaze to the Australian island state – one of only three World Heritage–listed temperate wilderness areas in the Southern Hemisphere (2015: 338). This chapter identifies that this 'ceaseless dance' takes a variety of guises in reporting environmental protest action and for our purposes, climate crisis. High-profile media stunts that the environmental movement is well-practised at delivering offer immediate news media coverage and exposure for the issue at hand but can often result in subsequent scepticism from journalists. At its worst, and as demonstrated in the Franklin Dam blockade, this attempt to manipulate media agendas can backfire on the movement, forcing 'neutral' journalists into a corner whereby they feel obliged to provide space to climate change sceptics in order to 'balance' the coverage. Political theorists such as Machin (2013) allude to the difficulty the environmental movement faces in walking a fine line between encouraging local, individual environmental 'actions' at the risk of losing sight of much bigger structural and political-economic changes. Social movement authors such as Hestres (2014) point to more contemporary strategies – primarily used by Internet-based climate change activist groups – which see campaigns targeted at a much narrower 'issues public', that is, people already engaged in an issue who may simply require information and a trigger to be mobilized. Christians *et al.* clarify that it is unlikely the dominant media are capable of supporting larger, structural change; it is only in outlets performing a radical media role that such fundamental challenge to power structures is enunciated and supported. For the most part, dominant media are attempting 'neutrally' to observe and report, with occasional incursions to facilitate action and civic engagement in the issues.

The arguments presented here suggest rough terrain for journalists attempting to cover climate crisis in the manner that the issues might demand. Certainly, while immediate gains and exposure for particular campaigns can successfully be achieved through the established media strategies of the environmental movement, the sustained impact of these 'stunts' is questioned. New shoots may be evident in the digital era, and there is some emerging evidence for this. The next chapter addresses challenges and options facing journalists attempting to successfully cover climate crisis. Existing models – Peace Journalism and Civic Journalism – are considered as we seek modes of comprehensive, accurate and truthful reporting of climate crisis which not only inform but activate the citizenry and, through this, potentially influence politicians and policy-makers.

References

Anderson, A. and Gabor, I. (1993) 'The yellowing of the greens', *British Journalism Review* 4(2): 49–53.

Atton, C. (2002) *Alternative Media*, London: Sage.

——— (2007) 'A brief history: The web and interactive media', in K. Coyer, T. Dowmunt and A. Fountain (eds), *The Alternative Media Handbook*, New York and London: Routledge, pp. 59–66.

Beck, U. (2009) *World at Risk*, Cambridge, UK: Polity.

Blühdorn, I. (2007) 'Sustaining the unsustainable: Symbolic politics and the politics of simulation', *Environmental Politics* 16(2): 251–75.

——— (2013) 'The governance of unsustainability: Ecology and democracy after the postdemocratic turn', *Environmental Politics* 22(1): 16–36.

Castells, M. (2000) *The Rise of the Network Society*, 2nd edn, Oxford, UK: Blackwell.

——— (2004) *The Power of Identity*, 2nd edn, Oxford, UK: Blackwell.

Chapman, G., Kumar, K., Fraser, C. and Gaber, I. (1997) *Environmentalism and the Mass Media: The North/South Divide*, New York: Routledge.

Christians, C., Glasser, T., McQuail, D., Nordenstreng, K. and White, R.A. (2009) *Normative Theories of the Media: Journalism in Democratic Societies*, Urbana and Chicago: University of Illinois Press.

Commission on Freedom of the Press (Hutchins Commission) (1947) *A Free and Responsible Press*, Chicago: University of Chicago.

Cottle, S. (2006) *Mediatized Conflict: Developments in Media and Conflict Studies*, Maidenhead, UK: Open University Press.

—— (2011) 'Taking global crises in the news seriously: Notes from the dark side of globalization', *Global Media and Communication* 7(2): 77–95.

Couldry, N. and Curran, J. (2003) 'The paradox of media power', in N. Couldry and J. Curran (eds), *Contesting Media Power: Alternative Media in a Networked World*, Lanham, MD: Rowman and Littlefield, pp. 3–15.

DeLuca, K. (2011) 'Environmental movement media', in J.D.H. Downing (ed), *Encyclopedia of Social Movement Media*, Thousand Oaks, CA: Sage, pp. 172–8.

Foxwell-Norton, K. (2015) 'Community and alternative media: Prospects for twenty-first-century environmental issues', in C. Atton (ed), *The Routledge Companion to Alternative and Community Media*, London and New York: Routledge, pp. 389–99.

Gamson, W.A. and Wolfsfeld, G. (1993) 'Movements and media as interacting systems', *Annals of the American Academy of Political and Social Science* 528: 114–25.

Hansen, A. (1993) 'Greenpeace and press coverage of environmental issues', in A. Hansen (ed), *The Mass Media and Environmental Issues*, Leicester, UK: Leicester University Press, pp. 150–78.

—— (2010) *Environment, Media and Communication*, London and New York: Routledge.

Hay, P. (1991–92) 'Destabilising Tasmanian politics: The key role of the Greens', *Bulletin of the Centre for Tasmanian Historical Studies* 3(2): 60–70.

Hestres, L. (2014) 'Preaching to the choir: Internet-mediated advocacy, issue public mobilization and climate change', *New Media & Society* 16(2): 323–39.

Hutchins, B. and Lester, L. (2006) 'Environmental protest and tap-dancing with the media in the information age', *Media, Culture & Society* 28(3): 433–51.

—— (2015) 'Theorizing the enactment of mediatized environmental conflict', *International Communication Gazette* 77(4): 337–58.

Juris, J. (2005) 'The new digital media and activist networking within anti-corporate globalization movements', *The Annals of the American Academy of Political and Social Science* 597: 189–208.

Kenis, A. and Lievens, M. (2014) 'Searching for "the political" in environmental politics', *Environmental Politics* 23(4): 521–48.

Lasswell, H.D. (1948) *The Structure and Function of Communication in Society*, New York: Harper & Bros.

Lester, L. and Cottle, S. (2009) 'Visualizing climate change: Television news and ecological citizenship', *International Journal of Communication* 3: 920–36. Accessed at http://ijoc.org/ojs/index.php/ijoc/article/view/509/371.

Machin, A. (2013) *Negotiating Climate Change: Radical Democracy and the Illusion of Consensus*, London: Zed Books.

Painter, J. (2011) *Poles Apart: The International Reporting of Climate Skepticism*, London: Sage.

Pantti, M., Wahl-Jorgensen, K. and Cottle, S. (2012) *Disasters and the Media*, London: Peter Lang.

Solesbury, W. (1976) 'The environmental agenda: An illustration of how situations may become political issues and issues may demand responses from government, or how they may not', *Public Administration* 54(4): 379–97.

Chapter 4

From frames to paradigms

Civic journalism, peace journalism and alternative media

Robert A. Hackett

The research reviewed in Chapter Two suggests that the key environmental deficit of conventional media is not the shortage of information but rather the failure to engage and motivate publics to participate effectively in collective solutions for climate crisis. Indeed, the Vancouver-based Climate Justice Project's focus groups, comprising environmentally concerned but politically disengaged citizens, support that assertion. We found that the main roadblock to participation was not lack of information but rather alienation from government, parties and other political institutions and from political activism, other people and 'the public'. Arguably, such subjectivity is promoted by currently hegemonic neoliberalism – a deep distrust of politics and collective action, the privileging of acquisitive individualism and the private sphere and, at worst, a sense of limitless entitlement to self-benefit regardless of the public good or corrective criticism – the personality type colloquially known as 'the asshole' that philosopher Aaron James (2012) fears is becoming more frequent in contemporary capitalism. Yet neoliberalism has not entirely vanquished the concern for connection, meaning and larger-than-self issues such as environmental well-being. Our focus groups, the interviews with environmental communicators and the research literature all suggest that certain approaches can indeed engage their audiences.

In particular, Chapter Two identifies frames and approaches that take news media beyond the monitorial role of information provision to help engage, motivate and mobilize concerned publics: frames that evoke a sense of efficacy and active citizenship; climate justice frames that highlight the voices of the marginalized and of global warming's primary victims; frames that make links between ethics and politics; frames that offer critical distance from a culture of luxury emissions, consumerism, narrow self-interest, the global North's development path, even of capitalism; frames that cultivate self-transcending altruistic values and subjectivity (more warriors, fewer assholes) and trust in the public; and frames that normalize collective political action and have a local register.

What is meant by 'framing' when it applies to news media? The concept is multi-dimensional. Frames are characteristic of individual cognitive processes (how we handle information), of stories and texts and of the process of selecting and constructing news stories. Frames denote 'principles of selection, emphasis, and presentation composed of little tacit theories about what exists, what happens and what matters' (Gitlin 1980: 6) and also what is good or bad and what is related to what. Framing is both a selective process – some aspects of the world are foregrounded, others marginalized or excluded – and an exercise of power. The ability to define a political issue such as climate policy through frames that resonate with publics and help them 'fix' its meaning is halfway to winning the policy battle. And the resources to establish politically favourable frames are very unevenly distributed in neoliberal society.

Unfortunately, environmentally productive frames do not always dance well to the drumbeats of dominant media. One roadblock is professional journalism's 'regime of objectivity' discussed in the Introduction. Journalism traditionalists will question the notion that journalism should be governed by extrinsic purposes or have predetermined storylines or intentional policy impacts. Such objections presuppose that conventional journalism is politically neutral and ideologically inert (a view encapsulated by a short promo for CBC's national television news program; its best-known newscaster, Peter Mansbridge, asserts that 'I don't make the news, I report it'). They overlook journalism's structured linkages to other institutions, the inevitability of framing as a narrative practice (stories cannot be told without implicit assumptions about what is relevant and important), the power relations embedded in framing practices and the predictable biases inherent in the conventional practices or 'strategic rituals' (Tuchman 1978) of objectivity. Such biases include elites over grassroots peacemakers or social change agents, events over processes or conditions and two-sided conflict over alternative perspectives and win-win solutions (Lynch and McGoldrick 2005a). As previously noted, the practice of 'balance' between two sides – in this case, climate scientists and deniers – distorted the American public's understandings of global warming for years (Boykoff and Boykoff 2004).

Moreover, aspects of the political economy of conventional news media – ownership or financial links to extractivist industries and to consumer culture via advertising and data mining and the disinvestment in journalism by conglomerate owners – are not conducive to explanatory, investigative and solutions-oriented journalism that would challenge politics and business as usual.

How, then, can more productive environmental frames be nourished within the media system? We can suggest three levels of potential change in climate journalism: best practices, paradigms and structures. These levels roughly parallel the three levels of journalistic agency identified by Berglez (2011) vis-à-vis the challenge of reporting the climate crisis: inside strategies that fine-tune conventional media logic (style and framing); outside strategies that introduce elements of external discourses (such as science or environmental communication); and 'beyond' strategies that would transform the whole way of thinking and doing journalism.

'Best practices' for climate journalism

Hire more science-trained reporters. . . . Don't compartmentalize climate news, spread it like butter to different topics. . . . Don't always trumpet 'climate change' in the headlines, introduce it more subtly. . . .

Journalism reviews, scholarly journals and NGO websites in recent years have abounded with advice on tweaking climate news within existing media. Helpfully, Bourassa *et al.* (2013) have identified eight thematic areas in the literature on 'best practices'. The first seven themes refer to reforms of conventional journalism, as follows:

1   Abilities. Training in effective communication of science and risk to the public, ethical retention of objectivity in the face of pressures from various sources spinning self-serving information, and sufficient knowledge of environmental issues are some of the prominent themes. Potential practical steps include workshops, professional internships and more collaboration between journalists, scientists and the academy.

2   Variability. Journalists need continually to learn about new topics and to otherwise adapt to the broadness of the beat and its overlap with other beats. Climate change is no longer a story about the reliability of scientific data but reaches into 'international affairs, food, mainstream politics, farming, transport, health, energy, taxation issues, and more' (Smith 2005: 1471).

3   Range of information. How can journalists provide context and avoid becoming conduits for vested interests' public relations? Should journalists include solutions, overcoming the unease amongst traditionalists with slipping from analysis to advocacy?

4   Sources. Diversity and decentralization are the motifs here. Don't over-rely on government officials; use more ordinary people and scientists, but translate them into a public idiom. In the words of a UNESCO guidebook for African journalists:

> Quote varied voices. Climate change affects everyone and everyone can respond to it in a different way. Think about both gender and generation. Climate change will affect men and women in different ways. Young people and old people are both more vulnerable than healthy middle-aged people. They also have different perspectives. Very old people have long memories and can describe decades of change. Young people will inherit the problems of climate change and so may have powerful perspectives.
>
> (Corcoran *et al.* 2013)

5   Balance and objectivity. Writers in this vein have accepted that a fabricated balance between climate science and opinion was detrimental to public understanding. It generated stories that were overly politicized and insufficiently factual. Such writers worry that framing news through preconceived

dichotomies (environment versus economy, etc.) may intensify polarization rather than cooperative solutions. (However, Chapter Two suggests that certain kinds of conflict framing can encourage public engagement and positive change.)

6   Newsworthiness. Climate change poses particular challenges in attracting audience interest. Stories that are complex, long-range, involve rationality or cooperation or consequences that are uncertain or in the future do not mesh well with conventional news values. There is also the risk of 'climate fatigue' if repeated too often.

7   Storytelling methods. Skilled narrative techniques can help address the challenge of low newsworthiness. Proffered recipes include using different angles; tying stories to interesting peoples, places and topics; reporting on solutions; highlighting local voices, human interest and 'real people', while unwaveringly presenting the big picture and not oversimplifying issues; differentiating local and global causes of weather-related disasters; explaining adaptation imperatives and options (Fahn 2009; Kirby and Radford 2011).

While many of these proposals are worthy, they face important limitations. Some of them, such as solutions-oriented reporting, could encounter resistance from traditionalists defending 'objectivity' (Hackett and Zhao 1998). Others, such as greater scientific training for journalists or creating more dedicated environment beats, would require greater resources at a time when major media corporations are disinvesting in journalism (Freedman 2010; McChesney and Nichols 2010). And arguably, some of these practices would yield little payoff in terms of audience engagement with climate politics; for instance, providing journalists with greater scientific training could be counterproductive in terms of relating to audiences (Dunwoody 2004).

How many of these practices would promote the re-orientation towards the world, the fundamental shifts of perspective and cultural narrative that environmental crisis necessitates? Some of them would shift us in a productive direction: more solutions; more context, local voices, human interest. But they do not necessarily generate fundamentally new frames or narratives, let alone the building blocks of more ecologically attuned subjectivity.

Some of these practices also run against the conventional media's grain, making greater demands on audiences (attention to complexity, willingness to challenge conventional wisdom) than are easily compatible with the commercial imperative of grabbing audience attention as cheaply as possible.

Moreover, a sustained shift in journalism requires training, editorial encouragement, organizational commitment and investments of time and money. Neither the traditional practices of conventional media nor the spontaneous outpouring of 'social media' (better termed 'digital connective networks' [Downing 2015: 108]) are likely to generate a fundamental green shift in journalism.

Appropriately, Bourassa's eighth theme encompasses options outside of traditional reporting methods, such as 'sustainable journalism' that would combine diligent research, precise language and fair reporting (traditional journalism desiderata) with educating the audience; initiating public discussion of environmental issues; actively seeking more environmental information; making corporations more accountable for their environmental record; and presenting the audience with possible solutions. This takes us into the terrain of 'beyond' strategies and alternative paradigms.

Is climate crisis journalism a new paradigm?

In calling for new paradigms, are we suggesting that news should become propaganda in the service of environmental causes? That the media field should be governed by logics external to itself?

Of course not. There is an irreducible difference between propaganda, which entails manipulating information for predetermined goals, and journalism, whose capacity for an independent gaze arguably helped constitute modernity (Calcutt and Hammond 2011). The point, rather, is that journalism is inherently a value-laden and political exercise. Its scanning of the world is driven by assumptions and routines that are related to power, and any systematically practiced paradigm has fairly predictable consequences. The regime of objectivity, for example, tends to over-access official sources and to induce political cynicism or passivity. The 'challenger paradigms' discussed in this chapter do not produce propaganda; rather, they start with an explicit value orientation (to public life and to non-violent conflict resolution) that informs practitioners' initial questions, sources and methods but does not lead to predetermined conclusions. While in this chapter I questioned the practices that too often pass for journalistic objectivity, the core promise of journalism – timely truth-telling in the public interest – is more relevant than ever in the context of global crisis. A line by 1960s troubadour Bob Dylan could be an epigram for global crisis journalism: 'so let us not talk falsely now, the hour is getting late.'

Broader than individual story frames, a journalism paradigm comprises integrated elements that usually include distinct philosophical groundings, an analysis of how media work and a repertoire of methods. Is climate crisis–oriented journalism a distinct paradigm in this sense? Certainly not at present, though there have been initial efforts at defining one (see 'sustainable journalism', mentioned previously). Should it become one? We have argued that climate action frames and environmental communication have multiple strands; similarly, there is no one-size-fits-all climate journalism. But extrapolating from the preceding chapters, we can suggest some of its most potentially fruitful features.

As noted in Chapter One, we follow American scholar Robert Cox (2007: 15–16) in recommending a *crisis orientation*, one that enhances society's ability to respond appropriately to environmental signals, makes information and decision-making processes 'transparent and accessible' to the public, helps equalize the resources for participation in environmental decision-making, and critically evaluates ecologically harmful policies and practices. If transposed to journalism, these criteria could help bridge the 'objectivity' standards of conventional reporting and the 'advocacy' work of alarmed citizens. Their adoption implies a recognition that journalism is an inherently political practice, that there are already established models of engaged or advocacy journalism and that nevertheless certain precautions would be needed – for example, avoid evaluating journalism through the single-minded lens of its environmental consequences. Cox's criteria are also reasonably consistent with the recognized monitorial and facilitative functions of journalism – holding power accountable, surveying the physical and social environment for threats to well-being and promoting inclusive societal conversation on matters of public importance (Hackett *et al*. 2013: 36). Insofar as Cox's model calls for a redistribution of discursive resources, it also stretches towards the radical role.

In addition to a crisis orientation, effective climate journalism would celebrate political action by ordinary citizens and tell success stories about climate politics to counteract cynicism and counterbalance routine reporting of the failures of conventional politics. The focus would be strongly local: How is our community connected with the causes, impacts and solutions for climate change? Journalism could normalize political engagement as something that ordinary people can and do undertake, provide more concrete 'procedural knowledge' about how to take political action and promote agency, particularly of people already concerned about climate change (Cross *et al*. 2015). Conflict can be presented in ways that evoke outrage and mobilization rather than paralysis and cynicism by using the classic movement-building tactics of identifying grievances, enemies, allies and solutions, which entail a willingness and capacity to undertake advocacy journalism. A prerequisite for this role is structural independence from governments and industries that are currently locked into a high-emission extractivist economy. Such journalism cultivates values and proffers frames, such as climate justice, that are likely to promote pro-environmental outlooks and behaviour. It transcends the monitorial reporting of media stunts and particular events, to explore causes, processes, consequences and potential responses to climate chaos.

Environmental news in the digital environment

The digital media environment obviously creates new tools, platforms and opportunities for environmental journalism. (Some of these are discussed in the case study in Chapter Seven.) Here, we just highlight a few developments. Researcher James Painter (Burke 2015) argues that digital media are 'disrupting' climate change coverage in several ways. The arrival of the millennial generation's 'digital natives' is providing a new audience for environmental news as an issue of particular concern to them, an audience that has powered the growth of outlets such as BuzzFeed, Vice and Huffington Post. High-quality niche sites like InsideClimateNews.org offer specialist information on climate change for journalists and other attentive publics. And technological advances enhance the portrayal of climate change science through better graphics and visuals and the capacity to analyze huge data sets.

More techno-buzz surrounds the potential use of virtual reality technology, like electronic headsets to create a three-dimensional interactive environment (Schmidt 2016). It's an extension of 'immersion journalism', which focuses on conveying experience rather than a narrative. Perhaps the virtual sensory experience, however safe and temporary, of floods in Bangladesh or droughts in Africa could induce privileged global Northerners to empathize with climate change victims. More likely, though, virtual reality will be used mainly for military and commercial purposes, selling everything from snowmobiles to houses.

It is too early to evaluate the transformative potential of such innovations. They are techniques rather than broad new paradigms, and as argued in Chapter Seven, we should question the utopianism that greets every new digital technology.

The digital mediascape also poses new challenges to long-established paradigms. Public service broadcasting has particular potential relevance to climate crisis communication. Historically, the rationale or mandate for public broadcasters such as ABC, BBC and CBC in Australia, the U.K. and Canada, respectively, has included the democratic roles of informing citizens, highlighting and amplifying public issues, facilitating discussion and communicative exchange between different segments of society, providing 'universally' available broadcasting service throughout the country, and reducing the impact of social inequalities on participation in public discourse. These are classic functions of the democratic public sphere, but they are not well fulfilled by entertainment- and profit-oriented commercial broadcasters. Additionally, in a country such as Canada, facing the enormous cultural, economic and media presence of the U.S., public broadcasting has been tasked with providing a communicative space that parallels the regional and national political systems, better enabling Canadians to address and hear one another.

Comparative research in the U.S. and five European countries has shown that public broadcasting makes a political difference (Aalberg and Curran 2012: 189–99). It is more likely to offer substantive public affairs news at time slots available to large audiences and to frame the news thematically (contextually) rather than episodically (focusing on disjointed and fragmented events – a complaint about news media voiced by the environmental communication practitioners interviewed for Chapter Five). Citizens of countries with relatively strong public broadcasters had, on average, higher levels of political knowledge and trust and confidence in politics. Impressionistically, it seems that while thoughtful environmental programming is a small portion of the overall schedule, public broadcasters are more likely than commercial networks to offer it. In Canada, CBC has broadcast programs such as *The Nature of Things*, hosted by geneticist and environmentalist David Suzuki, as well as *This Changes Everything*, Avi Lewis and Naomi Klein's critical documentary on the politics of climate change.

Like public broadcasters elsewhere, the CBC has faced threats to its funding and political independence from unsympathetic neoliberal governments. The digital mediascape poses additional challenges. The online cornucopia fragments audiences, segments them increasingly into like-minded opinion tribes and overrides national boundaries. It is more difficult to constitute broad regional or national publics that parallel the geographical scope of formal political representation or to attract people to 'serious' public affairs news. At the same time, the Internet presents new opportunities for collaboration, interactivity and distribution. Media reform organizations in various countries, from New Zealand and Taiwan to Canada and the U.K., are actively campaigning for the reinvigoration and 're-imagining' of public broadcasters' democratic mandate.

In the remainder of this chapter, though, I focus on what climate crisis communicators could learn from two lesser-known paradigms or 'corrective journalisms' (Cottle 2009). A full interrogation of each paradigm might examine its view of journalism's core purpose, its epistemological assumptions about journalism as a form of knowledge, its distinct concepts, its associated practices and institutions, its historical conditions of existence, its supportive or oppositional relation to mainstream journalism and broader power relations, its promoters and opponents, and its antagonisms or synergies with other paradigms (Hackett 2011).

I do not address all those questions here, but rather I sketch two approaches particularly relevant to climate crisis criteria, such as urgency, comprehensibility, local relevance, public engagement, support for political and social change, and agency, reflexivity and creativity for journalists. Civic (or Public) Journalism and Peace Journalism each have generated a theoretical rationale, methodological toolkit, some degree of empirical evaluation and a number of on-the-ground experiments. Each has anchorage in an intellectual discipline – deliberative democratic theory and peace and conflict studies, respectively – that parallels climate crisis journalism's mooring in environmental communication. Attention now turns to them.

Civic Journalism anyone?

The Public *or* Civic Journalism (CJ) movement gained traction in some U.S. newspapers during the 1990s. It too was born from a sense of crisis – not environmental, but political, a widespread sense of democratic 'malaise' and a 'disconnect' between American publics, on the one hand, and politicians and media, on the other. Its core premise is the ethical and practical requirement for journalism to not simply reproduce official political agendas but to actively help reinvigorate public life (see e.g. Merritt 1995; Rosen 1991). During the 1990s, newspaper editor Davis 'Buzz' Merritt and New York University academic Jay Rosen mapped out this new model in the context of declining newspaper circulations and growing popular distrust of the press. As a leading student of Civic Journalism put it, the 'overarching goal of increasing citizen participation in democratic processes' was accomplished by

> focusing attention on issues of concern to citizens, reporting on those issues from the perspectives of citizens rather than politicians, experts and other elite actors, offering citizens opportunities to articulate and debate their

opinions on issues, elaborating on what citizens can do to address those issues, organizing sites for citizen deliberation and action such as roundtables, community forums and local civic organizations, and following up on citizen initiatives through ongoing and sustained coverage.

(Haas 2004: 118)

Philosophically, CJ is clearly linked to the model of deliberative/public sphere democracy discussed in Chapter One, explicitly drawing upon such theorists as Habermas and Dewey. CJ exemplifies media's facilitative role. In emphasizing broad citizen participation in both revitalizing democracy and addressing social problems, CJ fits remarkably well with environmental communicators' call for public engagement in a crisis situation. Participatory approaches may well be better suited than traditional 'objective' journalism to the social-political conditions and uncertainties of late modernity (Howarth 2012).

But how well does it work in practice? Did it fundamentally alter news practices, audiences' civic engagement or political outcomes? Some observers think so. As arguably 'the most significant reform movement in American journalism' in the past century (Dzur 2002), CJ has 'worked its way into the fabric of American newsrooms' (Cross 2002). A study for the Pew Center for Civic Journalism, reviewing 651 projects, touted improved communities' public deliberative processes, favourable responses from audiences, improved skills of citizens, better journalistic practices, responsive policy outcomes, expanded number of civic organizations and broader inclusion of citizen voices in the news (Friedland and Nichols 2002).

But other assessments are more cautious. CJ's effectiveness is partly related to institutional venue. Small-town newspapers' urgent incentives to increase circulation and their tradition of community engagement helped make them CJ's strongest supporters. Scope matters too; a study of a daily paper in Savannah, Georgia, found that CJ was more evident in the routines and presentation of community news than in news about larger events and issues (Nip 2008) – a cautionary note for its application to global warming. Local television news content was less affected by CJ experiments (Maier and Potter 2001), perhaps because of commercial television's entertainment orientation and its higher per-story capital costs.

A review of research on three dimensions of CJ's impact suggests nuanced conclusions (Massey and Haas 2002). First, has CJ influenced journalists' professional attitudes? Surveys suggest American journalists are relatively more comfortable with CJ's less activist and more traditional shadings, such as using reporting to help communities move towards solutions to public problems (p. 564). While other surveys showed affinity for even CJ's more activist aspects, such as media-sponsored town meetings, journalists evaluated CJ by pragmatic rather than ideological criteria (pp. 564–65). CJ has not really convinced American professional journalists of its greater utility and of its critique of traditional journalism (pp. 565–66). Arguably, this reflects in part the strength of traditional journalism's regime of objectivity. Traditionalists question whether CJ-influenced newsrooms have transgressed the monitorial role by publicizing events that they themselves have arranged. (This allegation of conflicting interests begs the question of whether media can ever be simply detached observers.)

Second, did CJ change newsroom practices? It shifted political coverage away from process-oriented emphasis on the horserace (polls and strategies) and towards issues – more so in newspapers than television. But it had less success in shifting sourcing practices. Average citizens appeared more frequently in the news, but often simply as illustrative props to community problems and solutions that continued to be defined by elite sources (officials, politicians, accredited experts, police) (Massey 1998). An 'economic concern among for-profit news organizations with keeping information-gathering costs low' reinforced reliance on such sources (Massey and Haas 2002: 577).

Editors at 'civic' newspapers were more likely than counterparts elsewhere to say their organizations reported on solutions, convened community meetings and provided readers with reporters' contact co-ordinates (p. 568). But otherwise, traditional storytelling devices, such as the emphasis on conflict, prevail as much in civic as traditional newsrooms.

Third, did CJ succeed in changing people's civic attitudes and behaviours? Surveys suggest a positive impact on attitudes towards politics and civic life and knowledge of election-year issues and candidates' strategies as well as increasing voter turnout, interpersonal civic discussion and participation in community groups. On the other hand, CJ projects that focused on encouraging only deliberation did not necessarily lead to greater participation in collective problem-solving. Following from the analysis in Chapter Two, perhaps the missing link was the representation of such civic activism as a social norm, something that ordinary people are actually doing.

To be sure, CJ can be seen as a precursor of more recent forms of interactive and participatory online journalism (Schaffer 2015). One strand is 'public-powered journalism', in which media audiences collaborate with professionals to develop stories. A relevant example is iSeeChange – an independent platform on which ordinary folks can post their observations about local impacts of climate change – founded by Julia Kumari Drapkin, a multimedia producer with background in public and community media (Koski 2015).

But what is CJ's staying power within corporate media? The Pew Center's review of 651 CJ projects concluded that 'almost 20%' continued their CJ involvement for 'more than four years' (Friedland and Nichols 2002). But is the glass one-fifth full, or four-fifths empty? It seems that without the impetus of continued foundation funding, many for-profit newsrooms abandoned their commitment to CJ relatively quickly. The dominant media's coverage of the 2016 U.S. election campaign bears little resemblance to the deliberative democracy CJ was intended to promote. Reports of polls, tactics and Donald Trump's antics overwhelmed exploration of policy issues and options. The punditocracy's bafflement at the populist appeal of both Trump and Democratic candidate Bernie Sanders indicated a continuing gap between elite media and wide segments of the public.

What has been CJ's impact on public policy outcomes or on political culture beyond particular cities? Such impact would be difficult to isolate from a nexus of other factors – political campaign financing, electoral competitiveness, personal and demographic variables – and there is apparently little research on this question. Nor does there appear to be research to verify claims by some of its supporters that CJ could increase circulation and revenues during the 1990s, which may help account for why it has petered out in the dominant media.

CJ was hampered by the political economy of traditional news organizations; even those committed to CJ. Kurpius (2002) found that TV news departments' commitment to CJ was 'often trumped by traditional business concerns' (cited in Massey and Haas 2002: 569). A leading researcher, sympathetic to CJ's philosophy, argues that corporate media's political and commercial needs place distinct limits on their dabbling in CJ. They do not engage in advocacy of particular solutions, for example, and partner mainly with 'politically benign organizations' such as foundations or universities rather than political parties or unions. CJ offers only limited 'mobilizing information' to help citizens engage politically; it will promote voluntary community organizations without considering whether the issue is systemic or requires political action and policy change at regional, national or even international levels. Journalists still set the media agenda, making CJ a 'conservative reform movement' which is not institutionally accountable to the publics that it calls forth (Haas 2004). For more robust forms of civic journalism, argues Haas, turn to alternative media.

From her research on journalistic practices and motivations in Australia's alternative press, Forde (2011: 11–14) reached similar conclusions: community broadcasters, not commercial media, fostered the most successful forms of public journalism. The blockages in corporate media include the ethos of objectivity, the politically and economically convenient dependence on official sources, the potential financial cost of CJ projects and, perhaps most fundamentally, the incompatibility of the kind of consumerist subjectivity cultivated by commercial media with that of the politically active citizen. Consumerism reinforces individual rather than collective identities and action, which is at odds with building the political capacity to address climate crisis. It ideologically lubricates environmentally harmful 'lifestyles' and embeds the inequalities of a 'market economy'. While democratic citizens are in principle equal, consumers are not; their agency depends on their purchasing power. If the logic of democracy is one person, one vote, the logic of markets is one dollar, one vote. As that logic plays out in commercial, advertising-dependent media, it trends towards greater attention to the issues and sensitivities of affluent consumers than those of the less well-heeled. Some CJ projects did amplify issues of the poor, like a joint newspaper/television station initiative to uncover the roots of crime and involve poor, mainly black neighbourhoods in finding and implementing community-level solutions in Charlotte, North Carolina (Schaffer 2015). But the very fact that this initiative was nominated for a Pulitzer Prize indicates its relative rarity. More common are the class-related blind spots (labour issues, economic inequality, corporate crime) in the news agenda identified by the monitoring project NewsWatch (Hackett and Gruneau 2000).

A former journalist puts it pithily:

> I am experienced enough to know that . . . 'the bottom line' is still the decid-ing factor in what reporters do and do not cover. Journalists who are fortunate enough to work in profitable organizations offering a popularly accepted prod-uct tend to have a lot of free rein to do the kinds of 'community focused' work they want to do. Journalists working under economic constraints, or in strug-gling entities, or who are working for managers who think the route to good community journalism runs through the ad sales department are not so free.
>
> (Swanson 2001: 492–3)

Besides inattention to the political economy of corporate media, CJ arguably 'car-ries with it the limits of communicative democratic theory' (Compton 2000: 450). Here, Machin's (2013) critique of deliberative democracy, noted in Chapter One, rings true. In common with Habermas, CJ's view of communicative rationality is abstracted from people's embodied, historically situated experience, their rooted-ness in social contexts of class and gender. Such divorce from social context and power relations leads CJ to overestimate the possibility of consensus. 'Nor is an apparent consensus inherently desirable, if it means ratifying an unjust status quo or precluding further debate' (Hackett and Zhao 1998: 205).

Two proponents of CJ address precisely this criticism. Rather than refute it, they suggest ways that journalism could overcome it by nurturing

> a public sphere understood as comprising multiple discursive domains, in which different social groups can deliberate among themselves before doing so jointly. . . . Thus, journalists should aim to promote genuine participa-tory parity in the public sphere by ensuring that subordinate social groups enjoy the same opportunities as dominant social groups to articulate their concerns. Journalists can do this by foregrounding the issues of subordinate social groups and emphasizing salient social inequalities, and by offering citizens opportunities to reflect on . . . how [their] social locations affect their sense of problems and solutions.
>
> (Haas and Steiner 2006: 246)

Just so! This position essentially parallels Fraser's (1997) concept of counter-public spheres that provide space for subaltern groups to form their own identities and strategies before re-engaging with the broader one. But such space is not likely to be abundant in commercial, corporate news organizations. As we discuss later in the chapter, democratic engagement may find more fertile soil in alterna-tive media.

While its viability within conventional media is doubtful, Civic Journal-ism has left important legacies for climate crisis communication. It opened up debate about journalism's democratic purposes and its relationship with those it claims to serve. It de-naturalized organizational routines and orthodoxies, particularly objectivity. It invited journalists to be more reflexive about their practices and impact, especially the place of the public in their stories and the frames and master narratives employed (Compton 2000: 455). Its experiments, within their limits, showed a considerable potential for both journalists and publics to recover a sense of political agency. A similar impetus has motivated Peace Journalism.

Peace journalism

Given the ambiguous impact of Civic Journalism, we turn to a more recent paradigm that might resonate with our focus groups and environmental communicators alienated by conventional climate politics news. Briefly, as outlined by two of its leading exponents, Peace Journalism (PJ) is an analytical method for evaluating reportage of conflicts, a set of practices and ethical norms that journalism could employ to improve itself, and a rallying call for change (Lynch and McGoldrick 2005b: 270). In sum, PJ's public philosophy 'is when journalists make choices – of what stories to report and about how to report them – that create opportunities for society at large to consider and value non-violent responses to conflict' (Lynch and McGoldrick 2005a: 5).

PJ draws upon the insights of conflict analysis to look beyond the overt violence which is often tantamount to War Journalism. PJ calls attention to the context of attitudes, behaviour and contradictions. If War Journalism presents conflict as a tug-of-war between two parties in which one side's gain is the other's loss, PJ invites journalists to re-frame conflict as a cat's cradle of relationships between multiple stakeholders; to distinguish between stated demands, and underlying needs and objectives; to move beyond official sources to include other voices – particularly victims and those working for creative and non-violent solutions; to explore ways of transforming and transcending the hardened lines of conflict; and to report aggression and casualties on all sides, avoiding demonizing language and the conflict-escalating trap of emphasizing 'our' victims and 'their' atrocities. PJ looks beyond overt bloodshed to include other forms of everyday violence that may underlie conflict situations: structural violence; the institutionalized barriers to human dignity and well-being, such as racism; and cultural violence, the glorification of battles, wars and military power (Hackett 2006).

Israeli scholar Dov Shinar (2007: 200) offers a concise summary of PJ prescriptions for better journalism:

> Exploring backgrounds and contexts of conflict formation and presenting causes and options on every side so as to portray conflict in realistic terms, transparent to the audience
>> Giving voice to the views of all rival parties
>> Offering creative ideas for conflict resolution, peacemaking and peacekeeping

Exposing lies, cover-up attempts and culprits on all sides and revealing excesses committed by and suffering inflicted on people of all parties

Paying attention to peace stories and post-war developments more than the regular coverage of conflict

Shinar then bids caution and realism regarding both the prospects for implementing PJ in journalism practice and its impact in conflict situations. I return to the question of implementation in a subsequent discussion, after first considering whether the PJ model can be transposed to climate crisis journalism.

Peace journalism for climate crisis?

In the search for journalism adequate to the scale and urgency of climate crisis, PJ offers a growing repertoire of philosophical support, methodological guidelines and field experience from which to draw. In this section, I briefly thematize some of the potential affinities between PJ and climate journalism.

Although PJ's purpose is to reduce violent conflict rather than ecological destruction, its ethical horizon – a peaceful, just and sustainable global society – resonates with climate journalism. Both approaches aim to transform journalism into journalism that can transform the broader culture. Media reform is not only about media reform. But most PJ advocates also respect journalism's autonomy and the need for professional ethics and standards. PJ seeks news media that are more independent of established power, that are not suborned to propaganda from vested interests – including advocacy groups. As Lynch has put it:

> peace journalism is an advocacy position vis-à-vis journalism itself, but *it is not trying to turn journalism into something else.* If 'society at large' is provided with such opportunities [to value non-violent conflict resolution], but chooses not to take them, then there is nothing else journalism can do about it, while remaining journalism.
>
> (2008: 3–4; emphasis in original)

That 'something else', presumably, is propaganda on behalf of any particular organization. PJ retains a profound commitment to truth-telling in the public interest, but:

> On the other hand, there is no concomitant commitment to ensuring that violent responses get a fair hearing. They can take care of themselves, because the reporting conventions (still) dominant in most places, most of the time, ensure that they seldom struggle for a place on the agenda.
>
> (Lynch 2008: 4)

For similar reasons, climate crisis journalists need not make special efforts to grant access to climate science denialists or to extol the virtues of consumerism, economic growth or public cynicism about collective action. The biases of conventional news will normally reinforce those values effortlessly.

PJ finds intellectual anchorage in an academic discipline – peace and conflict studies, with particular reference to the pioneering work of Johann Galtung (Lynch and McGoldrick 2005a). The efforts to translate this anchorage into journalistic practice could inspire parallel work to link environmental communication and reporting practices in the pursuit of journalism that is both scientifically informed and politically empowering.

PJ and environmental communication scholars alike maintain that news reporting is neither ideologically neutral nor separate and detached from the 'events' that it reports. Interpretive frames necessarily influence the apparently neutral reporting of events. Peace journalists Lynch and McGoldrick (2005b) hypothesize a 'feedback loop' between journalism and political actions, arguing that conventional conflict reporting (which they regard as tantamount to War Journalism) creates incentives for conflict escalation and 'security crackdowns'. Environmental journalism scholars Boykoff and Boykoff (2004) argue that inappropriate 'balance' between science and opinion confused American public opinion for years. PJ enjoins self-reflexivity on the part of journalists vis-à-vis both the influences on and the predictable consequences of their own routine practices.

Scholars such as Lynch and McGoldrick recognize limits to journalism's power, given media organizations' unavoidable imbrication with broader social relations and political institutions, and yet seek to recover a sense of agency for journalists, resisting reductionist conceptions of the news as merely putty in the hands of powerful elites.

PJ's practices have much to offer climate journalism. Peace journalists broaden the range of sources and voices in the news beyond officials and technocratic experts to grassroots activists, solution-builders and the victims of war – a democratized pattern of access that resonates with climate justice. Peace journalists have found ways to expand the news agenda beyond today's events and to tell engaging narratives about contexts such as patterns of structural and cultural violence, the historical development of attitudes and policies by the parties in conflict, creative ideas for peaceful conflict resolution, processes of peace-building during and after conflicts, and the 'invisible' costs of war beyond bloodshed and destruction. The growing news attention and public recognition of soldiers' post-traumatic stress as a cost of war is an example of how journalism can render visible the previously unseen. Insofar as crisis-oriented climate journalism would extend the news agenda beyond protests and disasters such as oil spills to explore global warming's systemic roots, there are lessons to be learned from Peace Journalism's theory and practice.

Like Climate Justice Journalism, PJ seeks to transform relationships with audiences, or at least to evoke a different response. Preliminary evidence in Mexico, the Philippines, Australia and South Africa suggests that by contrast with conventional war reporting, PJ framing does generate (at least amongst focus groups in experimental settings) greater degrees of empathy, hope and cognitive engagement with counter-hegemonic arguments vis-à-vis war propaganda (Lynch 2014; McGoldrick and Lynch 2014). While it remains to be demonstrated on a broader scale, PJ's apparent impact is consistent with the public empowerment and larger-than-self values called for by environmental communicators.

Finally, both PJ and emergent climate journalism challenge conventional journalistic practices and self-understandings. They are inherently controversial and can expect to be ignored, dismissed or critiqued by journalistic traditionalists, some academics and (to the extent that such transformative journalisms gain traction) the powerful interests that would be less able to dominate news agendas. Advocates and practitioners of crisis- and engagement-oriented climate journalism could be forearmed by reviewing debates since the emergence of PJ in an annual journalism summer school in the U.K. in the mid-1990s (Lynch 2008: xi). German scholar Thomas Hanitzsch (2004a, 2004b) has been an especially prolific snowball-thrower. He argues that PJ wrongly assumes that journalism routinely overemphasizes violence, assumes an outdated view of media effects as powerful and linear, and adopts a naïvely realist epistemology, expecting news to provide 'truth' rather than 'distortion'. PJ inappropriately assigns journalism peacemaking tasks that are better suited to other institutions, says Hanitzsch, and in so doing, compromises journalists' integrity and neutrality.

PJ advocates have responded by clarifying misconceptions (they do *not* favour suppressing news that could jeopardize the prospects of peaceful outcomes, nor do they expect journalism alone to save the world), modifying positions (PJ aims to expose propaganda, but does not naïvely expect to provide unassailable 'truths') and, above all, continuing to problematize conventional 'objective' reporting practices as complicit in the escalation of conflict (see e.g. Lynch 2008). PJ aims to provide a journalism that is actually more complete, informative and truthful than conventional journalism and that can be justified in terms of the latter's own stated ideals.

Many of PJ's arguments, frames and practices could be transposed to crisis-oriented climate journalism. There are, however, important contrasts between the two journalisms.

'The war is on!': Paradigm disjunctures

It's June 18, 2014. Canada's federal government has just announced its long-expected support for the proposed Northern Gateway pipeline, one that would slash from Alberta's tar sands through First Nations territory in northern British Columbia to coastal ports. The reaction is swift and well-publicized. At a rally outside CBC headquarters in Vancouver, in front of television cameras and a thousand energized supporters, Grand Chief Stewart Phillip declares, 'The war is on!' Rousing cheers and street dances ensue (Prystupa 2014).

That 'war' metaphor has important implications for Peace Journalism's relevance to climate crisis. What if Naomi Klein is correct: 'Indigenous rights – if aggressively backed by court challenges, direct action, and mass movements demanding that they be respected – may now represent the most powerful barriers protecting all of us from a future of climate chaos' (2014: 380). Is it possible that, contrary to the precepts of PJ, saving the planet may require taking sides and escalating conflict in order to disrupt an ecocidal status quo?

Indeed, it could even be argued that in a state of planetary emergency, a more appropriate model might be the openly patriotic press of the allied powers during World War II, engaged in a life-and-death struggle against fascism. Defeats as well as victories were reported, but there was no pretence of neutrality. How might such wartime journalism be relevant to climate crisis – a sense of urgency, the sheer amount of coverage, the weaving of discrete news events into an overarching narrative, the identification of enemies and the framing of news as us-versus-them? On the other hand, wartime journalism implies censored and slanted news, the suppression of dissent in favour of unity against a common foe and a huge buy-in to journalism's collaborative role. But collaboration with whom? In wartime, this is collaboration with the government and the military. It is difficult to see those institutions as allies if ecological sustainability requires radical change.

Still, the question of advocacy journalism in relation to the agonistic politics of climate change hints at some important disjunctures between PJ and Climate Crisis Journalism (CCJ). Their definition of the core problem differs. In its dominant versions, PJ sees conflict itself, and the threat of conflict escalation to the point of violence, as the key issue – not any particular party to the conflict. CCJ would focus on global warming and its impacts on the human and 'natural' worlds and the (in)adequacy of societal and political responses. To mobilize effective responses, it may be necessary to bring millions of people who will not take no for an answer into the streets (Monbiot 2009), escalating conflict in order to challenge business as usual.

Likewise, they differ regarding the key shortcoming of journalism. For PJ, journalism too often contributes to conflict escalation and fails to convey the accurate and complete accounts of conflicts that notionally democratic societies need as a basis for informed policy. CCJ sees a range of environmental deficits in hegemonic media, above all their imbrication with consumerist culture and corporate capitalism. It is an open question whether these respective diagnoses point strategically in the same direction. PJ seeks to change journalism practices and representations so as to increase the likelihood of peaceful conflict resolution and make it less likely that news media contribute to conflict escalation; it calls for avoiding 'demonizing' one party to a conflict or identifying it as the enemy.

CCJ could well contribute to broadening the scope of conflict as a means of achieving social change (a strategy well understood in social movement practice); and in calling for increased analysis and attention to the causes of global warming, it could well lead to identifying the fossil fuel sector or other particular interests as targets for political action.

This approach parallels the struggles, alluded to previously, of Indigenous peoples on the front lines of resistance to extractivist capitalism. In Vancouver, anti-pipeline protesters, Aboriginal and settler allies alike, wear T-shirts emblazoned with 'Warrior up!' The 'warrior' concept is arguably a 'trope', a figure whose meaning differs between discourses. At one level, it is a colonial stereotype emerging from settler society, alongside 'drunken Indian' and 'noble savage', for example. In recent decades, however, it has been re-appropriated by some Indigenous nations defending their homelands from settler-controlled development (such as the expansion of a golf course onto sacred Indigenous lands, resulting in an infamous standoff at Oka, Quebec, in 1990). It has particular recent relevance in the context of territorial defence against resource extraction and energy mega-projects and thus, climate change. Within Indigenous nations where it has been deployed, the concept can be a divisive one, particularly when it is taken to connote violence and the identification of enemies. There appears to be more consensus when 'warrior' is associated with sacrifice on behalf of others, rootedness in the community and customary laws of their people, collective self-defence against external threats, resistance to colonialism, and a spiritual and ethical struggle that can be politicized through 'self-transformation and self-defence against the insidious forms of control that the state and capitalism use to shape lives according to their needs – to fear, to obey, to consume' (Alfred 2005: 29).

Just as the warrior concept is ambiguous, so too are its implications for Peace Journalism. On the one hand, even if Indigenous warriors are committed to non-violence, the concept does entail taking sides and assigning blame, identifying the colonizing state, developers and extractivist companies as aggressors. On the other hand, in the Kanien'keha language of the Kanehsata:ke ('Oka') community, the word for warrior is 'Rotiskenrakeh:te', usually translated as 'those who carry the burden of peace' (Gabriel 2014). Parenthetically, these considerations point to the interplay of media discourses and practices with subjectivity (Corner 2011); if the previously discussed 'asshole' is a byproduct and bulwark of neo-liberalism, the warrior may be an oppositional antidote. But the valorization of the warrior does not seem to be part of Peace Journalism's normative framework.

Still, the contrast between PJ and CCJ should not be exaggerated. In part, this is because neither PJ nor CCJ are monolithic paradigms. Both have 'mainstream' or 'liberal' versus 'radical' variants. We have already noted environmental communicators' differing views on conflict framing. A roughly parallel dichotomy is relevant in PJ debates. The co-editor of the British peace movement journal *Peace News* argues that while it has usually followed PJ's dominant practices, it has sometimes found it necessary to 'assign blame' in conflict situations (Rai 2010: 220). British journalism educator Richard Keeble (2010: 63–4) argues for 'a radical political re-theorising of journalism and more specifically peace journalism' as an 'essentially political practice'. He critiques the 'dominant strand' in PJ as focusing too narrowly on reforming professional routines rather than on campaigning/advocacy journalism, alternative/oppositional media and the fresh possibilities for participatory and citizens' journalism through the Internet.

Thus, affinities between PJ and CCJ are more pronounced by comparing their respective liberal and radical versions. Liberals seek reforms within existing media and policy institutions (e.g. PJ as 'better' journalism rather than a fundamental challenge to its procedures and self-understandings). They pursue consensus and dialogue, based on the assumption that underlying interests (as distinct from stated demands) are ultimately compatible, that war and ecological degradation are unintended consequences in nobody's interests – a position parallel to deliberative democracy. Radicals are more likely to adopt a view of society as characterized by fundamental antagonisms and by governing logics that however ultimately destructive they be, they can only be challenged and reversed through resistance to identifiable enemies and the formation of counter-hegemonic alliances. In that perspective, PJ is relevant insofar as it provides discursive resources (such as structural contexts and propaganda critiques) that support struggles for social change.

Can we get there from here?

The liberal variants of both paradigms assume and seek change within the field of professional journalism. Lynch is concerned to recover a sense of agency for journalists, one that is missing in the radical functionalism of some theories of the media, such as the Propaganda Model (Herman and Chomsky 2002). PJ's aim to challenge the inevitability of War Journalism framing is commendable. But the skills and resources needed for either Peace or Climate Crisis Journalism (especially if it is informed by a Climate Justice meta-frame) do not mesh well with the constraints, discussed in our Introduction, imposed by conventional media – particularly, ownership disinvestment in news, the continued national bases (and biases) of media organizations and audiences, and structural ties to consumerism and capitalism. Peace Journalism seems to have flourished only under certain conditions, such as societies where media contributed to destructive internal conflict, and/or news organizations with a stake in avoiding their audiences' dissolution into opposing camps, and/or societies emerging from authoritarian rule, where journalism's professional norms may be relatively open to self-reflexive change (Lynch 2010). Similarly, as noted in the Introduction, CCJ is likely to find some market and institutional conditions more conducive than others.

Nor should we assume that the digital media environment automatically bypasses the blockages of hegemonic media. Previously, we discussed some of the new opportunities for online independent journalism and popular mobilization (see also Chapter Seven). Yet the commercialized Internet and 'social media' are also complicit in the spread of disinformation and misinformation, the segmentation of users into like-minded opinion tribes, the growing precarity of journalistic labour, the erosion of professionalization and the profusion of entertaining clickbait. Well-resourced and highly skilled journalists remain as essential as ever in covering an issue as complex as climate crisis. Unfortunately, it seems that in the Western corporate media, journalists have neither sufficient incentives nor autonomy vis-à-vis their employers to transform the way news is done without support from powerful external allies (Hackett 2006, 2011: 45). We need to consider alternative media as an emerging site for Climate Crisis Journalism.

Alternative media

Unlike Peace and Civic Journalism, alternative media are often taken to comprise not so much a coherent paradigm as a residual category: media that are 'alternative' to whatever media dominate the cultural and political 'mainstream'. In at least one sense, alternative journalism within such media is inherently political in that it is 'always a reminder of what the dominant forces in society *are not providing, or are not able to provide*' (Forde 2011: 45; emphasis in original). As we discussed in the Introduction, the vitality in the practice and study of the field is reflected in the wide range of labels it has received and debates around key conceptual questions: Should alternative media be defined by what they differ from ('mainstream' media) or rather by more positive characteristics, and if so, which ones? Should their distinctiveness be understood simply as divergence from a dominant model or as opposition to it? What form of domination is the object of such opposition, and are oppositional 'alternatives' necessarily 'progressive' in the broad sense of seeking a fairer distribution of social, economic, cultural and political resources (Couldry 2003, 2010; Hackett 2011: 46–7)?

This chapter bypasses those debates in order to compare recognized characteristics of alternative media with what our research suggests about effective Climate Crisis Journalism. One way to distinguish alternative media is via their distinctive 'logics'. In uneven and sometimes contradictory ways, an alternative medium manifests one or more of the following driving forces: a logic of participation (self-representation, access to the means of media production, capacity building for communities and subordinate groups); counter public formation and facilitation, which registers both the facilitative and radical roles of media identified by Christians *et al.* (2009) in forming deliberative and potentially oppositional publics; critical-emancipatory logics, which highlight counter-hegemonic content and form; and heterodox-creative logics of cultural radicalism and aesthetic innovation (Mowbray 2015). Mowbray's schema is especially useful in opening the possibility that alterity is not a monopoly of particular media institutions or people; alternative practices can sometimes appear in unexpected places.

Focusing more specifically on alternative *journalism*, Atton and Hamilton (2008) identify these aspects:

- Oppositional or counter-hegemonic content (alternative frames; coverage of issues, events and perspectives that are marginalized or ignored by hegemonic media; criteria of newsworthiness that emphasize the threats that the established order poses to subaltern groups, rather than vice versa)
- Participatory production processes; horizontal communication both within news organizations and with readers and audiences – communicative relationships that both reduce the gap between producers and users and empower ordinary people to engage in public discourse
- Mobilization-orientation; a positive orientation towards progressive social change and productive connections with (but not subordination to) social movements

- Localism and engagement with communities, whether these be defined in terms of shared locale or shared interests
- Independence from state and corporate control and from commercial imperatives; individual or cooperative ownership
- Low degree of capitalization, often reliant on volunteer labour, grants and donations; conventional production values and standards of professionalism may be less important.

While low capitalization limits alternative media's reach, most of the other characteristics mesh well with Climate Crisis Journalism that would seek to inform and mobilize counterpublics, engage local communities and challenge entrenched power. Yet, as we have noted, there has been relatively little previous attention to the role of alternative media in climate change communication. An early study found surprisingly few differences between New Zealand's major national daily and an online alternative outlet (Kenix 2008). But such a conclusion was probably due to contingent factors. Mainstream media attention to climate issues temporarily spiked during the research period, following reports by the Intergovernmental Panel on Climate Change and Al Gore's documentary *An Inconvenient Truth,* likely de-differentiating alternative and mainstream news agendas. Moreover, the study focused on climate science rather than climate politics and selected media outlets and hypotheses that did not necessarily register key mainstream/alternative distinctions. The study is nevertheless a useful reminder not to reify such distinctions, as commercialism and globalizing journalistic norms generate ongoing pressures for convergence (Kenix 2011).

Nevertheless, more recent research suggests tenacious differences. Gunster (2011) compared how mainstream and alternative media in British Columbia reported upon the December 2009 climate negotiations in Copenhagen. Alternative media, he concluded, offered more optimistic and engaged visions of climate politics than the cynical, pessimistic and largely spectatorial accounts dominating conventional news. While alternative media were deeply critical of the spectacular failure of 'politics-as-usual' at that summit, they invited responses of outrage and (collective) action rather than (individualized) despair. Informed by a deeper, more sophisticated and sympathetic understanding of diverse climate activisms, alternative media (re)positioned political action as viable forms of agency vis-à-vis climate change. Such portrayals

> not only feed upon the hope, nourished by historical example and consciousness, that democratic pressure can compel [existing] institutions to behave differently, but also awaken the political imagination to the utopian prospect of inventing new institutions and . . . forms of politics in response to environmental crisis. Such an expansion of the conceptual and affective spaces for climate politics produces an orientation that is simultaneously more critical and pessimistic about the limits of existing structures and practices, yet also

more optimistic about the opportunities for collective political agency and intervention.

(Gunster 2011: 492–3)

In a companion year-long overview of alternative media's climate change coverage, Gunster (2012) argued that this most optimistic disposition was often due to such media's consistent attention to inspirational stories of political success – concrete examples of civic engagement, political struggle, innovative and effective public policy and transformative change in communities, institutions and governments that sustain and invigorate feelings of hope.

Can the practices and culture of alternative media be scaled up to provide more space for a challenger paradigm such as Climate Crisis Journalism? Some of its methods can be adopted in traditional news organizations. On the whole, though, the challenger paradigms have been marginalized, orphaned (like Civic Journalism) or co-opted (like citizens' journalism transformed into 'user-generated content' for the corporate press). The 'beyonder' strategy of a wholesale shift in journalistic modes and frames that requires collective, well-resourced and organized efforts within journalism education and news institutions is not likely to occur within the structural and policy framework of hegemonic, market-driven media. It would require an exercise in 'public reformism', of concerted action to achieve intended change (Curran 2010), discussed in the Conclusion. Movements for ecological sustainability and for democratic media reform may have much in common, in part because the prospects for better climate journalism would be greatly improved by an institutional and policy environment that supports diversity and investment in independent media. The distinctive contributions of such media are explored in Chapters Six and Seven, but first we consider one key aspect of their practices and philosophical underpinnings – contending ways of framing conflict.

References

Aalberg, T. and Curran, J. (eds) (2012) *How Media Inform Democracy: A Comparative Approach*, New York and London: Routledge.

Alfred, T. (2005) *Wasáse: Indigenous Pathways of Action and Freedom*, Toronto: University of Toronto Press, Higher Education Division. Accessed at http://site.ebrary.com. proxy.lib.sfu.ca/lib/sfu/detail.action?docID=10116332.

Atton, C. and Hamilton, J. (2008) *Alternative Journalism*, London: Sage.

Berglez, P. (2011) 'Inside, outside, and beyond media logic: Journalistic creativity in climate reporting', *Media, Culture & Society* 33(3): 449–65.

Bourassa, E., Amend, E. and Secko, D.M. (2013) 'A thematic review and synthesis of best practices in environment journalism', *Journal of Professional Communication* 3(1): 39–65.

Boykoff, M.T. and Boykoff, J.M. (2004) 'Balance as bias: Global warming and the US prestige press', *Global Environmental Change* 14: 125–36.

Burke, K. (2015, October 22) 'Data journalism is disrupting climate change coverage (An interview with Dr. James Painter)'. Accessed at http://www.globaleditorsnetwork.org/press-room/news/2015/10/data-journalism-is-disrupting-climate-change-coverage/.

Calcutt, A. and Hammond, P. (2011) *Journalism Studies: A Critical Introduction*, London and New York: Routledge.

Christians, C., Glasser, T., McQuail, D., Nordenstreng, K. and White, R. (2009) *Normative Theories of the Media: Journalism in Democratic Societies*, Urbana and Chicago: University of Illinois Press.

Compton, J. (2000) 'Communicative politics and public journalism', *Journalism Studies* 1(3): 449–67.

Corcoran, T., Scherer, C., Shanahan, M. and Shubert, W. (2013) *Climate Change in Africa: A Guidebook for Journalists*, UNESCO Series on Journalism Education. Accessed at http://www.iaj.org.za/images/stories/toolkits/climate_change.pdf.

Corner, J. (2011) *Theorising Media: Power, Form and Subjectivity*, Manchester and New York: Manchester University Press.

Cottle, S. (2009) *Global Crisis Reporting: Journalism in the Global Age*, Maidenhead, UK: Open University Press.

Couldry, N. (2003) 'Beyond the hall of mirrors? Some theoretical reflections on the global contestation of media power', in N. Couldry and J. Curran (eds), *Contesting Media Power: Alternative Media in a Networked World*, Lanham, MD: Rowman & Littlefield, pp. 39–54.

——— (2010) 'Introduction to section I', in C. Rodriguez, D. Kidd and L. Stein (eds), *Making Our Media: Global Initiatives Toward a Democratic Public Sphere, Vol. 1, Creating New Communication Spaces*, Cresskill, NJ: Hampton Press, pp. 24–8.

Cox, R.J. (2007) 'Nature's "crisis disciplines": Does environmental communication have an ethical duty?', *Environmental Communication: A Journal of Nature and Culture* 1(1): 5–20. Accessed at http://www.informaworld.com.proxy.lib.sfu.ca/smpp/section?content=a778982174&fulltext=713240928.

Cross, A. (2002) 'A powerful experiment', *Quill* 90(5): 4. Accessed at http://proxy.lib.sfu.ca/login?url=http://search.ebscohost.com/login.aspx?direct=true&db=ufh&AN=6893469&site=ehost-live.

Cross, K., Gunster, S., Piotrowski, M. and Daub, S. (2015) *News Media and Climate Politics: Civic Engagement and Political Efficacy in a Climate of Reluctant Cynicism*, Vancouver, BC: Canadian Centre for Policy Alternatives.

Curran, J. (2010) 'The future of journalism', *Journalism Studies* 11(4): 464–76.

Downing, J.D.H. (2015) 'Conceptualizing social movement media: A fresh metaphor?', in C. Atton (ed), *The Routledge Companion to Alternative and Community Media*, New York: Routledge, pp. 100–10.

Dunwoody (2004) 'How valuable is formal science training to science journalists?', *Comunicação e Sociedade* 6: 75–87.

Dzur, A.W. (2002) 'Public journalism and deliberative democracy', *Polity* 34(3): 313–36. Accessed at http://www.jstor.org/discover/10.2307/3235394?sid=21105384079811&uid=3739400&uid=2129&uid=3737720&uid=70&uid=4&uid=2.

Fahn, J. (2009) 'Climate change: How to report the story of the century', *Science Development Network*. Accessed at http://www.scidev.net/global/climate-change/practical-guide/climate-change-how-to-report-thestory-of-the-cent.html.

Forde, S. (2011) *Challenging the News: The Journalism of Alternative and Community Media*, New York: Palgrave Macmillan.

Fraser, N. (1997) *Justice Interruptus*, New York: Routledge.

Freedman, D. (2010) 'The political economy of the "new" news environment', in N. Fenton (ed), *New Media, Old News: Journalism & Democracy in the Digital Age*, Los Angeles: Sage, pp. 35–50.

Friedland, L.A. and Nichols, S. (2002) *Measuring Civic Journalism's Progress: A Report Across a Decade of Activity*, Washington, DC: Pew Center for Civic Journalism. Accessed at http://www.pccj.org/doingcj/research/measuringcj.pdf.

Gabriel, E. (2014) 'Those who carry the burden of peace', *Voices Rising/Nations Rising* (April 8). Accessed at http://nationsrising.org/those-who-carry-the-burden-of-peace/.

Gitlin, T. (1980) *The Whole World Is Watching: Mass Media in the Making and Unmaking of the New Left*, Berkeley: University of California Press.

Gunster, S. (2011) 'Covering Copenhagen: Climate politics in B.C. media', *Canadian Journal of Communication* 36(3): 477–502.

—— (2012) 'Radical optimism: Expanding visions of climate politics in alternative media', in A. Carvalho and T.R. Peterson (eds), *Climate Change Politics: Communication and Public Engagement*, Amherst, NY: Cambria Press, pp. 239–67.

Haas, T. (2004) 'Alternative media, public journalism and the pursuit of democratization', *Journalism Studies* 5(1): 115–22.

Haas, T. and Steiner, L. (2006) 'Public journalism: A reply to critics', *Journalism* 7(2): 238–54.

Hackett, R.A. (2006) 'Is peace journalism possible? Three frameworks for assessing structure and agency in news media', *Conflict and Communication Online* 5(2). Accessed at http://www.cco.regener-online.de/2006_2/pdf/hackett.pdf.

—— (2011) 'New vistas for peace journalism: Alternative media and communication rights', in I.S. Shaw, J. Lynch and R.A. Hackett (eds), *Expanding Peace Journalism: Comparative and Critical Approaches*, Sydney: Sydney University Press, pp. 35–69.

Hackett, R.A., Gruneau, R., with Gutstein, D., Gibson, T. and NewsWatch Canada (2000) *The Missing News: Filters and Blind Spots in Canada's Press*, Ottawa and Toronto: Canadian Centre for Policy Alternatives/Garamond Press.

Hackett, R.A., Wylie, S. and Gurleyen, P. (2013) 'Enabling environments: Reflections on journalism and climate justice', *Ethical Space* 10(2–3): 34–46.

Hackett, R.A. and Zhao, Y. (1998) *Sustaining Democracy? Journalism and the Politics of Objectivity*, Toronto: Garamond [now University of Toronto Press].

Hanitzsch, T. (2004a) 'The peace journalism problem: Failure of news people – or failure on analysis?', in T. Hanitzsch, M. Loffelholz and R. Mustamu (eds), *Agents of Peace: Public Communication and Conflict Resolution in an Asian Setting*, Jakarta: Friedrich Ebert Stiftung, pp. 185–206.

—— (2004b) 'Journalists as peacekeeping force? Peace journalism and mass communication theory', *Journalism Studies* 5(4): 483–95.

Herman, E. and Chomsky, N. (2002) *Manufacturing Consent: The Political Economy of the Media*, New York: Pantheon.

Howarth, A. (2012) 'Participatory politics, environmental journalism, and newspaper campaigns', *Environmental Journalism* 13(2): 210–25. Accessed at http://dx.doi.org.proxy.lib.sfu.ca/10.1080/1461670X.2011.646398.

James, A. (2012) *Assholes: A Theory*, New York: Doubleday.

Keeble, R. (2010) 'Peace journalism as political practice: A new, radical look at the theory', in R.L. Keeble, J. Tulloch and F. Zollmann (eds), *Peace Journalism, War and Conflict Resolution*, New York: Peter Lang, pp. 49–67.

Kenix, L.J. (2008) 'Framing science: Climate change in the mainstream and alternative news of New Zealand', *Political Science* 60(1): 117–32.

—— (2011) *Alternative and Mainstream Media: The Converging Spectrum*, London and New York: Bloomsbury Academic.

Kirby, A. and Radford, T. (2011) *25 Tips for Climate Change Journalists*. Accessed at http://underthebanyan.wordpress.com/2011/08/16/25-tips-for-climate-change-journalists/.

Klein, N. (2014) *This Changes Everything: Capitalism vs the Climate*, Toronto: Knopf Canada.

Koski, O. (2015) 'How participatory journalism turns news consumers into collaborators', *Nieman Reports* (November 17). Accessed at http://niemanreports.org/articles/how-participatory-journalism-turns-news-consumers-into-collaborators/.

Kurpius, D.D. (2002) 'Sources and civic journalism: Changing patterns of reporting?', *Journalism & Mass Communication Quarterly* 79(4): 853–66. Accessed at http://doi.org/10.1177/107769900207900406 and http://jmq.sagepub.com.proxy.lib.sfu.ca/content/79/4/853.full.pdf+html.

Lynch, J. (2008) *Debates in Peace Journalism*, Sydney: Sydney University Press.

—— (2010) Personal interview, University of Sydney, June 25.

—— (2014) *A Global Standard for Reporting Conflict*, New York: Routledge.

Lynch, J. and McGoldrick, A. (2005a) *Peace Journalism*, Stroud, UK: Hawthorn.

—— (2005b) 'Peace journalism: A global dialog for democracy and democratic media', in R. Hackett and Y. Zhao (eds), *Democratizing Global Media: One World, Many Struggles*, Lanham, MD: Rowman & Littlefield, pp. 269–88.

Machin, A. (2013) *Negotiating Climate Change: Radical Democracy and the Illusion of Consensus*, London: Zed Books.

Maier, S.R. and Potter, D. (2001) 'Public journalism through the lens: How television broadcasters covered campaign 96', *Journal of Broadcasting & Electronic Media* 45(2): 320–34. Accessed at http://www.tandfonline.com.proxy.lib.sfu.ca/doi/abs/10.1207/s15506878jobem4502_8#.VeE46rxViko.

Massey, B.L. (1998) 'Civic journalism and nonelite sourcing: Making routine newswork of community connectedness', *Journalism & Mass Communication Quarterly* 75(2): 394–407. Accessed at http://proxy.lib.sfu.ca/login?url=http://search.ebscohost.com/login.aspx?direct=true&db=hus&AN=509720606&site=ehost-live.

Massey, B.L. and Haas, T. (2002) 'Does making journalism more public make a difference? A critical review of evaluative research on public journalism', *Journalism & Mass Communication Quarterly* 79(3): 559–86. Accessed at http://doi.org/10.1177/107769900207900303.

McChesney, R.W. and Nichols, J. (2010) *The Death and Life of American Journalism*, Philadelphia: Nation Books.

McGoldrick, A. and Lynch, J. (2014) 'Audience responses to peace journalism', *Journalism Studies*. Accessed at http://www.tandfonline.com/doi/abs/10.1080/1461670X.2014.992621?journalCode=rjos20.

Merritt, D. (1995) *Public Journalism and Public Life: Why Telling the News Is Not Enough*, Hillsdale, NJ: Lawrence Erlbaum Associates.

Monbiot, G. (2009) 'If you want to know who's to blame for Copenhagen, look to the U.S. Senate'. Accessed at http://www.theguardian.com/commentisfree/2009/dec/21/copenhagen-failure-us-senate-vested-interests.

Mowbray, M. (2015) 'Alternative logics? Parsing the literature on alternative media', in C. Atton (ed), *Routledge Companion to Alternative and Community Media*, London: Routledge, pp. 21–31.

Nip, J.Y.M. (2008) 'The last days of civic journalism', *Journalism Practice* 2(2): 179–96. Accessed at http://doi.org/10.1080/17512780801999352 and http://www.tandfonline.com.proxy.lib.sfu.ca/doi/abs/10.1080/17512780801999352#.VdqxMLxViko.

Prystupa, M. (2014) 'Vancouver crowd jubilant as "war" declared on Northern Gateway', *The Observer* (June 18). Accessed at http://www.vancouverobserver.com/news/grand-chief-jubilant-he-declares-war-northern-gateway?page=0,1.

Rai, M. (2010) 'Peace journalism in practice – *Peace News*: For non-violent revolution', in R.L. Keeble, J. Tulloch and F. Zollmann (eds), *Peace Journalism, War and Conflict Resolution*, New York: Peter Lang, pp. 207–21.

Rosen, J. (1991) 'Making journalism more public', *Communication* 12: 267–84.

Schaffer, J. (2015) 'If audience engagement is the goal, it's time to look back at the successes of civic journalism for answers'. Accessed at http://www.niemanlab.org/2015/06/if-audience-engagement-is-the-goal-its-time-to-look-back-at-the-successes-of-civic-journalism-for-answers/.

Schmidt, N. (2016) 'Full immersion', *Ryerson Review of Journalism*. Accessed at http://rrj.ca/full-immersion/.

Shinar, D. (2007) 'Peace journalism – the state of the art', in D. Shinar and W. Kempf (eds), *Peace Journalism – The State of the Art*, Berlin: Verlag Irena Regener, pp. 199–210.

Smith, J. (2005) 'Dangerous news: Media decision making about climate change risk', *Risk Analysis* 25(6): 1471–82. Accessed at http://onlinelibrary.wiley.com/doi/10.1111/j.1539-6924.2005.00693.x/abstract.

Swanson, D. (2001) 'Review of *Public Journalism and Political Knowledge* by A.J. Eksterowicz and R.N. Roberts', *Social Science Journal* 38(3): 491–3.

Tuchman, G. (1978) *Making News: A Study in the Construction of Reality*, New York: Free Press.

Chapter 5

Contesting conflict? Efficacy, advocacy and alternative media in British Columbia[1]

Shane Gunster

Closing the 'hope gap'? Alternative perspectives, alternative media

Summarizing the latest research examining news coverage of climate change (Hart and Feldman 2014; O'Neill *et al.* 2015), *Climate Central* blogger John Upton (2015) noted that 'news cycles tend to be dominated by horror and carnage – a recipe for depression that spills into climate change coverage, fueling what some experts call a "hope gap" that can lead people to fret about global warming but feel powerless to do anything about it.' Both U.S. and U.K. media, Upton explained, 'are struggling to produce stories about climate change in ways that are engaging for their audiences. Instead, they're fueling senses of hopelessness on the issue.' Commenting upon this trend, public opinion researcher Anthony Leiserowitz noted that 'we find in our audience research that even the alarmed [those most concerned about climate change] don't really know what they can do individually, or what we can do collectively. We call this loosely "the hope gap", and it's a serious problem. Perceived threat without efficacy of response is usually a recipe for disengagement or fatalism' (cited in Upton 2015).

Documenting the many shortcomings and gaps in commercial media coverage has become a cottage industry in the field of environmental communication. As recent reviews of the field have noted, much of this work has taken the form of critical quantitative and qualitative analysis of mainly Western, primarily print and online media texts (Olausson and Berglez 2014; Schäfer and Schlichting 2014). A smaller number of studies have ventured both upstream and downstream in the discursive flow of news, investigating the production of media content (e.g. Berglez 2011; Smith 2005) and its reception by audiences (e.g. Cross *et al.* 2015; Olausson 2011).

Much less attention, however, has been allocated to those who operate within the discursive terrain dominated by conventional news (and who are, therefore, intimately familiar with both its possibilities and its limitations) but who are also actively developing alternative communication and journalism practices to address and overcome the 'hope gap'. Scholars of independent and alternative media, for example, have long observed their potential not only to offer a critical perspective on dominant economic and political structures but also to incubate and nurture radical, alternative visions of democratic politics and to develop new models of participatory communication (Downing 2001; Forde 2011). Despite the affinities between critical perspectives in climate communication and the radical practices of alternative media, little academic work has investigated the potential such media hold to close the 'hope gap'.

In one of the few academic studies to address this question, I conducted a comparative analysis of how commercial and independent media in British Columbia (B.C.) reported the December 2009 climate negotiations (Gunster 2011). As noted in Chapter Four, independent media offered a much more engaged, optimistic and hopeful vision of climate politics than the cynical, pessimistic and largely spectatorial accounts which dominated commercial media. A companion study of a year's worth of independent media coverage of climate change–related issues (Gunster 2012) similarly documented a greater emphasis upon solutions and, in particular, stories of *political success* in which civic activism, combined with innovative approaches to public policy, were celebrated as concrete evidence of the possibility and benefits of collective, political action.

In this chapter, I build upon this research, reporting upon the results of 11 interviews with alternative media journalists, editors and publishers, and leading environmental activists and advocates and exploring their perspectives on the relationship among news, politics and public engagement with climate change. The participants work within a diverse array of B.C.-based alternative and independent media, political advocacy groups, environmental organizations and think tanks (Appendix). They play an active role in climate change communication, including the production of media content (e.g. news stories, opinion and editorial pieces, advocacy campaigns, etc.), serving as sources for journalists and reporters, engaging with stakeholders (e.g. industry, government, First Nations, local communities, etc.), and directly communicating with the public in online and offline venues.

British Columbia offers an ideal site for exploring the intersection of communication, efficacy and engagement around climate and energy politics. Over the past decade, the province has implemented a number of progressive environmental policies, including North America's first carbon tax as well as strict regulations mandating future electricity generation to be carbon neutral (Rhodes and Jaccard 2013). At the same time, the provincial government is a strong proponent of resource development, including plans for a massive expansion of natural gas mining for export to Asian markets (Chen and Gunster 2016). B.C. is home to a thriving environmental movement with a diverse range of organizations and groups (Salazar and Alper 2002) which have mobilized public opposition to 'neoliberal extractivism' (Fast 2014). Provincial First Nations, in particular, are leading local, grassroots resistance to a variety of projects such as two highly controversial proposals to build new pipelines to increase exports of bitumen from the Alberta tar sands (McCreary and Milligan 2014). Project proponents, including government politicians, businesses and some labour leaders, have aggressively championed resource development as a source of economic growth, government revenue and employment. These issues attract significant levels of public and media attention, with citizens deeply divided over how to balance the seemingly competing imperatives of environmental protection and economic growth.

Historically, British Columbia has suffered from one of the highest levels of media concentration in North America, with a single company exercising a virtual stranglehold over commercial daily newspapers and local broadcast news for much of the past two decades (Edge 2007). Such market dominance – coupled with the ideologically conservative, pro-business orientation of local commercial media – has stimulated demand for alternative sources of news, especially in the area of environmental and resource politics. In recent years, a number of independent journalists and organizations have emerged to fill the void and Vancouver in particular has developed a thriving ecosystem of alternative media. All of the interviewed participants occupy key nodes in these emerging and evolving networks of journalism, communication and advocacy. Their knowledge and experience afford them an ideal vantage point to reflect upon the relationship between news and public engagement as well as the particular ways in which their own media strategies and practices operate to narrow the gap between 'threat perception' and 'efficacy of response'. Interviews were conducted between October 2013 and November 2014, typically ranged from 45 to 90 minutes and covered a wide range of topics around communication, politics and climate change. All interviews were transcribed and qualitatively analyzed to identify key points of convergence and divergence in the opinions, experiences and approaches of our participants to these topics.

Given the intensely polarized nature of energy and climate politics in the province, all of the interviews spent considerable time exploring how our participants conceptualized *political conflict,* both as a ubiquitous feature of the discursive and cultural landscape and also as an element of their own communicative practice. While each participant offered a unique perspective on the relationship between conflict and civic engagement, two broad clusters of opinion coalesced around opposing assessments of the role that conflict can and should play in communicating with the public. For some, conflict frames primarily serve to alienate the public from climate and energy politics, portraying it as a space of partisan bickering, irreconcilable divisions and protracted gridlock. Getting people (re)engaged instead requires an emphasis upon concrete, pragmatic and often bipartisan solutions which only become visible when cooperation, compromise, dialogue and problem-solving displace well-worn cycles of conflict. Others, however, insist that narratives of conflict – especially in B.C., where the fight over neoliberal extractivism has become so deeply entrenched – are not only inescapable but also represent a valuable opportunity to get citizens involved in political action and struggle. These strategic and tactical differences rehearse the distinction (discussed more extensively in the Introduction) between the 'facilitative' role of journalism – in which news media strive to create and enhance a public sphere in which dialogue and deliberation open up new opportunities for the formation of political consensus – and a 'radical' role – where journalists prioritize identifying and challenging injustice – and then mobilizing citizens and supporting movements organized to oppose them.

Displacing conflict: Seeking solutions, investigating complexity, mobilizing influentials

Geoff Dembicki is the lead sustainability writer for *The Tyee*, an independent digital news magazine with a focus upon sustainability, politics and social justice which has rapidly become one of the leading sources for alternative news and opinion in British Columbia since it was founded in 2003.[2] Dembicki has written extensively about the politics of climate, environment and energy over the past decade. Media coverage of climate change, he suggested, has increased as the confrontational politics associated with carbon infrastructure projects has intensified, giving news a familiar template through which to represent the issue:

> [W]hile I think it's good that climate change is getting more attention, and protests against pieces of infrastructure are getting covered, and those critical voices are being brought into the mainstream media, the result is that the entire mainstream narrative around climate change is almost always defined by conflict. So it's one group fighting another, one country calling out another country for not achieving targets, and it results in this very pessimistic frame where it's hard to feel that anything you do can have a real impact.

Chris Wood, the coordinating editor with The Tyee Solutions Society – a non-profit partner which works with *The Tyee* to produce 'solutions-oriented journalism' (see discussion in this chapter) – agrees that ideological conflict has 'bedevilled' constructive engagement with climate change: 'There's this left-wing, anti-corporate, anti-globalization, anti-trade, anti-business, anti-profit nexus of ideas which embraces climate change; the folk who are for trade, profit, corporations, globalization . . . say, "Well, in that case, if those guys are for climate change, we're against it".' Journalism that enforces the divisions between these two warring ideological camps does little other than reinforce public scepticism about ever making progress.

For Dembicki, conflict narratives are not only a barrier to agency and hope, but they also lock key stakeholders and constituencies into a polarizing message track which prevents them from communicating with the public (and each other) in a thoughtful and constructive manner. Instead, their core communications objective becomes supplying news media with content that can be easily slotted into conventional journalistic formulas and, consequently, will generate media attention. While this might produce good copy, it ultimately marginalizes important issues from public discourse:

> During the recent B.C. election, I found it extremely difficult to get any traction on LNG [liquefied natural gas] – even though . . . it was public knowledge

that this was going to completely destroy B.C.'s climate targets. . . . But what I think is that people had been reading so much coverage of certain types of symbols, such as the tar sands or Enbridge [the Northern Gateway pipeline], and symbols framed in certain ways, that it was hard to get a sense of urgency about something like LNG because it just wasn't on people's radar. And I would ask environmental groups about it at press conferences, and they were not even that enthusiastic to talk about it, because it didn't fit with their objectives, with the frame they had put on climate change.

The rhetorical inertia of conflict narratives, compounded by their easy and productive articulation with dominant routines and patterns of news production, helps create a self-reinforcing cycle in which this simple form of storytelling displaces more complicated and less predictable accounts of climate change. It fortifies the 'prevailing view in the mainstream media and among the public that any progress on climate change is going to be fought over bitterly and will be decided through conflict' (Dembicki), which ends up producing a real 'blind spot' with respect to solutions that can emerge out of design, technology or policy innovation.

Representatives of advocacy groups also described the conflict frame as both pervasive and problematic. Kevin Sauvé is the communications lead in the B.C. office of the Pembina Institute, a Canadian think tank which specializes in environmental and energy policy. He argued that a reliance upon the 'usual suspects' in climate change stories ended up excluding a wider variety of sources who could help the public engage with climate change in novel and constructive ways:

Sources tend to be the same when it comes to . . . who's being represented: environmental groups pitted against politicians, environmental groups pitted against industry – often missing some very key players that are . . . at the forefront of climate. So, particularly, when it comes to things like adaptation, we're not looking at the people, the decision-makers, say, in communities that are actually taking the research from science and then using that research to make decisions on a daily basis about the work that they do in planning communities . . . that's a real missed opportunity in the mainstream media, because I think that these people stand to actually make climate change very tangible for everyday people.

Unfortunately, these adaptation stories cannot compete with the drama, emotion and 'good guys and bad guys' (Sauvé) which drive conflict narratives and, in turn, intensify polarization around climate change. P.J. Partington, a Toronto-based climate analyst also with the Pembina Institute, noted that 'conflict has served to . . . reinforce people's pre-existing opinions on the matter, and hasn't served to advance understanding in any way, because people just look at the piece, and they agree with whoever it is they already agreed with. So I think finding a way to present [climate science] that's less confrontational gives people the space to think about the issue more, and maybe learn something from it.'

David Beers is the founder, publisher and editor of *The Tyee*. Reflecting upon his own experiences working as a journalist in commercial media, he explained that there was 'this weird situation in newsrooms, where you had all these journalists that were trained in the ethos of investigation and finding things out for the common good, but their only mandate was to find out what was just going to hell, what sucked.' Not only was this a profound waste of the skills, capacity and mandate of journalists, but also it failed to address the needs of those audiences who want and need to learn about *solutions* – not simply out of idle curiosity, but to guide their own thinking about problem-solving. In response, Beers and others at *The Tyee* prioritize 'solutions-focused journalism' which shifts media's focus away from an obsessive but superficial fixation upon bad news (e.g. conflict, scandal, disaster, etc.) towards a sustained investigation into the diversity of solutions through which people and governments can begin to tackle problems which otherwise appear irresolvable. Identifying and diagnosing problems (such as climate change) is not, Beers suggested, what is missing from the news; rather, the problem is that's where most reporting tends to stop:

> At a certain point . . . everybody agrees there's a problem, and there's a pent-up anxiety and anxiousness to show the way. . . . [F]or there to be a rich and vibrant democratic conversation, you can't just give vent to people's angers and frustrations. You have to present people with a positive alternative, that together, if we can form a rough consensus, we can possibly streak towards. At which point, the conversation becomes really much more affirming. . . . That's just a lot different from everyone sitting around watching TV and saying . . . 'That sucks, all politicians are bad, scientists are evil.'

Instead, 'we offer our services as journalists to go about exploring and investigating what might be a solution to a problem that we all agree that we face.' And the value of such journalism is not simply measured by the hope that it injects into public discourses otherwise dominated by gridlock and pessimism. It also keeps faith with the basic mandate of news to provide an objective, accurate and useful description of the world in which we live. 'To really show people how policy works, how government works – how government, business, NGOs can work together to solve something – you really have to not just report on when it goes to hell. You've got to report on how it's going right, or might go right.'

Over the past several years, 'solutions journalism' has been attracting attention from journalists, news organizations and the general public, and it is increasingly positioned as a means to generate attention and engagement from people who are otherwise turned off politics. The Solutions Journalism Network (SJN), for example, was founded in 2013; it works with journalists and newsrooms to integrate a greater emphasis upon solutions in their work. One of the network's co-founders is David Bornstein, a *New York Times* journalist who authors a regular column entitled 'Fixes'. 'The feedback system known as journalism,' he explains, 'is based on the idea that the way to improve society is to show people where we're going wrong. It's like pointing out your children's mistakes every morning and expecting that this will make them better people. Children need examples. They need to know that different behaviour is possible and wins notice. Society needs the same thing' (cited in Rosenstiel 2015). Beyond circulating ideas about how to solve specific problems, the broader, cumulative and more transformative impact of such journalism is to challenge pervasive public cynicism about political engagement, collective action and the possibility of institutional change. 'When people feel their institutions are beyond repair and their fellow citizens are untrustworthy, they are less likely to think about engaging in the public sphere. They come to resign themselves to the status quo and focus on areas where they can assert control [such as in the private sphere, through consumption or lifestyle change]. That weakens democracy' (Bornstein cited in Rosenstiel 2015).

Preliminary survey research suggests that news content with a solutions focus does have a positive impact upon levels of efficacy and engagement as compared to coverage that does not prioritize solutions (Curry and Hammonds 2014; Curry *et al.* 2016). However, advocates such as Bornstein draw a sharp distinction between the rigorous, investigative and critical qualities of good solutions journalism and the facile, shallow species of 'feel-good' news which organizations such as *The Huffington Post* have recently embraced to drive traffic to their site. Serious journalism about solutions demands critical scrutiny of the claims of advocates and a detailed investigation of relevant evidence and thus requires both significant resources and organizational autonomy.

After *The Tyee* established itself as a credible source of alternative news and analysis (and was attracting a significant audience, especially amongst progressive opinion leaders), Beers was approached by charitable foundations seeking to fund sustained coverage of key issues which fit their research and advocacy mandates. Cautious about straying into the territory of public relations, Beers instead founded The Tyee Solutions Society (TSS), a non-profit organization which partners with others to fund 'catalytic journalism', 'solutions-oriented reporting that uses traditional investigative techniques to empower citizens with the information needed to seize opportunities for positive change. TSS reporting examines and explains the facts and identifies achievable solutions impacting the lives of Canadians' (Tyee Solutions Society 2015). As Beers explained, 'if an organization within its mandate needs to answer a question they can contract the Tyee Solutions Society to have journalists try to answer the question through the format of journalism.' Such work 'doesn't have a predetermined answer . . . it doesn't have an advocacy slogan attached to it. That's what journalists are good at doing, they're good at chasing questions.' As of April 2016, the TSS has co-produced 20 investigative series (each involving a dozen or more lengthy news stories) within four broad areas of focus: education and youth, energy, food security and income inequality. In addition to publishing the stories in *The Tyee*, the TSS works with other media partners to increase distribution of the reporting as well as through public engagement events which facilitate collaboration between citizens, policy-makers and elected officials. While these investigative pieces are distinct from the ongoing news and commentary produced by *The Tyee*, they have made a significant contribution to the organization's objective of empowering the public through an emphasis upon solutions.

Beers identified three different, but complementary, approaches through which solutions-focused journalism can 'catalyze concrete positive change.' The first is 'living the solution', in which individuals describe their own experiences with a particular form of behavioural, institutional or social and political change. Exemplary of this approach was the idea of the '100-mile diet': B.C. journalists Alisa Smith and James MacKinnon wrote in *The Tyee* about their attempt to subsist on a diet consisting only of food sourced within 100 miles of their home. The popularity of the online articles led to a best-selling book (Smith and MacKinnon 2007) and helped spawn a global movement devoted to dietary localization. Second, journalists can investigate and publicize innovative, local, small-scale experiments which are often highly successful but largely invisible to the broader public. In these cases, the analytical focus becomes questions of scale, reproducibility and barriers: If a specific practice, technology or policy is so effective, how can it be applied more broadly, and what social, economic or political barriers are preventing such expansion? Finally, Beers noted the value of exploring solutions in other jurisdictions, which are often neglected because of the parochial sensibilities (and shrinking resources for news-gathering) of commercial media. Learning about and from other places can shake up public acceptance of the status quo and enliven political debate about the full range of choices available to citizens and governments. How, for example, has Norway managed the development and governance of its energy resources (and the profits from them) compared with Canada? What might Vancouver learn from the cycling policies and infrastructure of Portland, Copenhagen or Amsterdam? While much of this information already exists in reports from NGOs and academic studies, the creativity and expertise of journalists *as storytellers*, combined with their ethical and professional commitments *as fact-checkers*, can give their reporting upon solutions a credibility and a rhetorical appeal which can get the public interested, engaged and excited.

Producing solutions-oriented content depends upon developing and sustaining journalistic capacity, an achievement of *The Tyee* that Beers was especially keen to emphasize. Steady employment for journalists is, he observed:

> no small thing these days. Especially work that allows you to have a beat, and write in a sophisticated way. . . . There's very few jobs like that left in journalism. So we've provided a home for that. And sometimes I think of ourselves as one of those monasteries during the Dark Ages . . . where journalists are in there making illuminated manuscripts and waiting for the Enlightenment. It's a really dark time for journalism, and somebody's got to keep doing beat-driven, in-depth, sophisticated journalism, just to keep it going. . . . I think we're building capacity among the journalistic set.

Beers noted the difficulties faced by Andrew Nikiforuk, a Calgary-based investigative journalist who has done a lot of work on the oil sands: 'The analysis that he came to was inconvenient for corporate media, which was that the oil sands are incredibly emissions-intensive, and are affecting the local ecology, the global ecology, and the national economy all in negative ways. Well, you start writing about that and you don't really have a home anymore.' The success of *The Tyee* allowed Beers to welcome Nikiforuk as the organization's first 'Energy and Equity' writer in residence in 2010–11, allowing him to 'widen his scope to examine all angles of Canada's morphing into a petro state' and 'push deeper beyond the day's fleeting headlines chronicling disasters and commodity price swings' (Beers 2010). In addition to writing dozens of articles and commentary pieces, Nikiforuk's residency led to the publication of *The Energy of Slaves* (2012), which used the analogy of slavery to explore how our largely invisible dependence upon oil has shaped our lifestyles, morality, economy and political system.

For Dembicki, the combination of journalistic autonomy and a focus upon solutions enables a shift beyond more simplistic conflict frames which pit one group against another. Such frames, he argued, are often symptomatic of a lack of journalistic resources, experience and knowledge, prevalent because they allow one to produce content easily, quickly and cheaply. Conversely, Dembicki's gradual accumulation of experience and knowledge on the energy/climate beat has allowed (and motivated) him to tell a much more sophisticated and, ultimately, hopeful set of stories about solutions:

> I've been . . . looking at the whole climate change debate as this . . . big system of interlocking parts that all depend on each other, and there are certain parts that are just locked together and are not moving, and that's when you look at the international climate change conferences, or our debate over coal or the tar sands or pipelines. And all the media attention is on those parts that aren't moving because they've somehow become imbued with the biggest political stakes, economic stakes. . . . And so I've tried to find all the parts that are moving quite a bit faster . . . the clean tech industry . . . China . . . designing [energy efficiency] IT software.

Especially important for both Dembicki and Beers are the identification of points of possible compromise and consensus – 'movable parts' – which are otherwise obscured by ongoing conflict between different stakeholders. One of the most surprising revelations for Dembicki in his series 'Greening the Oil Sands', for example, was the fact that 'the oil industry and environmental groups both support a carbon price, but they've never come together on the same stage and said, "Prime Minister Harper, this is something we all support." And the reason for that is because they're always just fighting all the time . . . because they're fighting so much [both groups] can never see where they have similar objectives.' For Beers, this was 'a huge story. It's one of the most censored stories ever. So finally

Bloomberg [New York–based international news agency] comes around and goes, "Ah, this can't be true"; phones up everybody and finds out it is true, and then they did their own story. . . . [F]ine-grained and credible journalists can be seen as honest brokers of that conversation in ways that activists and highly invested NGOs cannot.'

The Tyee's solutions emphasis was a perfect fit for the similar ambitions of many working within advocacy organizations to shift climate discourse away from conflict towards solutions. Sauvé, from the Pembina Institute, was especially enthusiastic about moving from 'he-said, she-said' frames to 'solutions-focused narratives'. He described, for example, an ongoing Pembina campaign entitled 'Green Energy Futures' featuring weekly profiles of:

> a community or a person . . . that's doing something related to renewable energy – that is already on the ground, this is happening, this is for real. This is something that is not pie-in-the-sky technology . . . it is grounded in practice. And it is not just scientists, it is everyday people that are either very innovative or very concerned or very enthusiastic . . . it's not just the story about politicians and environmental groups banging heads all the time.

Ben West is executive director of Tanker Free B.C. and, as one of the most prominent representatives of the environmental movement in the province, frequently appears in commercial media as a spokesperson around climate and energy issues. 'We are increasingly shifting our framing to be about alternatives,' he explained. 'So really showing choices . . . [providing] a pathway to reduced demand. And it is not pie-in-the-sky, unicorns and fairytales. It is talking about buses and trains and building codes and land use plans. . . . It's all pretty normal stuff that we're all familiar with, it's really just a matter of political choices, and by and large it's stuff that's overwhelmingly popular.'

Both West and Sauvé underlined the necessity for solutions-based discourse to become more concrete, practical and local to resonate with audiences. Otherwise, as West observed, 'too often the . . . alternative energy conversation is in these very broad brush strokes, and it isn't local/specific enough. . . . We'll fight a specific pipeline in a specific area, but then we'll talk about green energy all across the world without much more detail than that.' Awareness of concrete alternatives creates the space for conversation, deliberation and awakening our political imagination to new ways of organizing social life:

> In Canada, the clear narrative for most people is 'This is just the way things are', and I think we really need to get more pilot projects actually built that show that there are better ways of doing things, that really tie this back to climate change so that it's . . . top of mind for people when they have conversations and they think, 'Well, why aren't we talking about high speed rail like they're doing in Europe, and Asia, and Latin America, and all over the place . . . ?'
>
> (West)

People need to be able more easily to imagine themselves (and their children) employed in green jobs; their communities, cities and regions enjoying the benefits of more sustainable infrastructure; and their governments responsive to democratic participation in collective choices about the future. Solutions need to become as 'real' to the public as the carbon-intensive lifestyles, habits and infrastructure they will replace.

For some of those working within alternative media and advocacy organizations, there was a strong affinity between prioritizing solutions and conceptions of an ideal audience composed of motivated and empowered agents of change. By definition, alternative media do not address 'mass' audiences but instead tend to be directed towards individuals and communities which share a common set of progressive social values, including commitments to sustainability and social justice. Within this broad demographic, however, *The Tyee* especially prizes and targets 'influentials', that is, individuals who are 'in a position to make positive change' (Beers). This objective flows out of an understanding of social and political change as driven by politicians who are responsive to signals from influentials within their base:

> I would identify an influential as a twenty-two year old with three thousand Twitter followers. I would also identify an influential as a deputy minister, or a policy director at a think tank, or a mid-sized business operator, or a Native elder, or a school principal. Just somebody who basically is a nexus, stands at the centre or at the top of some kind of wider network. That the politician says, 'I don't see how I can survive without the support of that network.'
>
> (Beers)

For Beers, good information – and, in particular, new information about under-reported solutions – can, in the right hands, catalyze policy innovation. Ideally, revelation generates enthusiasm and excitement; solid journalistic practices supply sophistication and credibility, and, together, they encourage and enable influentials to champion solutions within their networks and institutions. 'Our job,' Beers summarized, 'is to act as a kind of pollinating bee, and move among people who are experts and influentials in an area, and see if through the act of journalism, you can actually catalyze some concrete positive change.'

Dembicki similarly identified engaging with opinion leaders – 'influential people in universities, media, business' – as a priority. 'My ideal reader by default is someone who's closely engaged on these issues: probably an academic or business person, or even another journalist. And those are the people who tend to have the most positive reactions to the type of reporting I do. Because it's tricky to focus so heavily on some of these moving parts . . . because they scatter people's pre-held notions about climate change.' He was especially pleased, for example, that the complexity of his work on 'greening' the oil sands allowed those actually working in the area to engage with the series in a constructive fashion. For Beers, attending to the genuine complexity of solutions requires abandoning, or at least tempering, the polemic stance which often dominates debates over climate and energy, and which might have more instinctive appeal to progressive audiences:

We could have come to the conclusion that the tar sands were just fundamentally wrong and destructive and therefore we had to throw everything we had at telling everybody every day just how bad they are, until someday finally we all agree that they were bad and they had to be shut down. But what are the odds? Any sane person would assess the amount of political and financial capital that's invested in the oil sands and say, '*The Tyee* saying that the oil sands are bad is not going to change anything.' So what if, instead, we put information out in the world that showed how the oil sands could be part of a transition. Well that's a complicated story to tell, right?

Media do not just produce content, they also produce audiences. For *The Tyee*, that means cultivating audiences with the deliberative capacity, intellectual appetite and political flexibility to appreciate, understand and engage with complexity and ambiguity and be open to having their views and assumptions challenged by different perspectives and new information. It is 'a gathering place where you can feel dignified and imagine that you're going to be enlightened . . . rather than just be pandered to, or have your buttons pushed' (Beers). One should not overemphasize this idea of an impartial 'solutions broker' frame: there is little question that an open and vigorous commitment to progressive values, politics and constituencies animates *The Tyee*'s journalism and editorial philosophy, distinguishing it from conventional news. Yet the focus upon enlightening influentials also suggests an approach to social change which prioritizes getting useful information into the right hands over more confrontational strategies of political mobilization and activism.

The Pembina's Sauvé shared a similar preference for targeted advocacy communication which can serve as a resource for decision-makers. 'We're very differentiated in the sense that we actually are looking for influencers. It's less now about this big sweeping public communication and raising alarm bells *en masse*, and more "who do we need to speak with, who do we need to talk to about this policy recommendation x and policy recommendation y?"' Mass communications strategies, Sauvé suggested, are having diminishing efficacy: first, the general public has been 'pushed' as far as it will go on the issue, and second, 'the more time I spend speaking to certain decision-makers, the less I am convinced that they're now swayed entirely by public thinking on climate change.' Scepticism about democratic accountability, in other words, means that scarce communications resources may be better deployed 'influencing the influencers, particularly for right now, because this is likely where we're going to find the most effective channel for moving the levers [of change].'

Embracing conflict: Dramatizing politics, mobilizing communities, normalizing activism

Is an emphasis upon conflict inevitably corrosive for efficacy, agency and hope? While all of our participants were critical of the predictable and formulaic patterns of conflict which dominate conventional news, some argued that conflict narratives are an inescapable and, in fact, essential part of good climate change communication. Conflict stories which intensify polarization, cultivate and focus outrage, and celebrate struggle can facilitate the transition from awareness and concern to political engagement and activism (Schwarze 2006). This perspective openly embraces a radical role for journalism in which the critique of power, inequality and injustice – and the empowerment of citizens and social movements – mediates a more traditional emphasis upon neutrality and balance in the portrayal of conflict.

Jamie Biggar, executive director with Leadnow – a progressive political advocacy organization which runs national, issue-based campaigns – explained that there are two kinds of archetypal narratives which have taken shape around climate politics. The first emphasizes the failure of politicians and traditional institutions to address climate change, represents key stakeholders as hopelessly gridlocked and positions the public as disgusted but helpless bystanders to dysfunctional processes. That is a very demotivating, disempowering story that leads to cynicism. And it is the story of climate politics which tends to drive conventional news agendas (Gunster 2011). However, 'there is another story in which institutional leaders are somewhat secondary, and what is actually primary is a fight between global fossil fuel companies and place-based, but global civil society' (Biggar; also see Klein 2014). This second story is not only more accurate:

> it is also much more empowering, because in the second story what you talk about or highlight is victories and defeats, but what you are highlighting is normal people who are getting involved, often successfully, against enormous odds. And that is really inspiring. And it fits with people's zeitgeist of the times, which is that things are really wrong in ways that are hard to articulate, and the levers of control of our society seem more and more distant. So where are there people who are taking things into their own hands and being successful? That is a much more motivating conflict story.

Indeed, Biggar argued that a 'huge part' of public engagement in climate change 'is figuring out what is an *accessible conflict* that you can get people into in order to challenge and hopefully transform what is going on. And, of course, the major answer to that is opposing pipeline projects, and other forms of dirty energy projects, where there is a physical, concrete thing on the ground that you can literally, physically stand up against and have a whole bunch of levers for trying to stop.' In other words, rather than dismiss *all* stories about conflict as alienating, we must distinguish between the *paralyzing, cynical* conflict frames recycled by conventional news and the *accessible, generative* and *mobilizing* conflict narratives favoured by social movements and activists.

As one who regularly oscillates between the worlds of journalism (as publisher, editor and writer for *The Common Sense Canadian*, an independent site which produces news and commentary about the relationship between environment and economy) and political advocacy (as a documentary filmmaker and anti-LNG activist), Damien Gillis offered a similar perspective on the relationship between conflict and engagement. The motif of conflict lies at the heart of his most recent film, *Fractured Land*, which profiles a young B.C. First Nations lawyer, Caleb Behn, battling the expansion of fossil fuel extraction in his community's traditional territory. Conflict within and between nations and communities, conflict within families, conflict within ourselves, conflict with nature: for Gillis, it is precisely the mediation of Behn's experiences through these themes that gives the film its power to engage audiences, as well as stimulate deeper reflections upon the culture, polity and economy in which we live. 'The conflict that I'm really talking about is the conflict of values and ideas, and it is natural for that to exist within an individual, and within a community, and within a country. It is healthy for people to be conflicted about the kind of big choices they are making in their life or as a society.' Invoking the concept of dialectic, Gillis argued that the clash of opposing ideas is often essential for social innovation and progress: 'people hunger for that, and it is very human to question ourselves and question these things, like a kitchen table debate.' He contrasted this with the 'conflict between two dogmatic positions' which tends to drive news formats, especially segments which feature political commentary and debate. 'How uninteresting, how predictable and unproductive and essentially useless that dialogue is; it is not even a dialogue, it is just two talking heads presenting previously contemplated and articulated opinions that have nothing to do with facts. There is no drama in that. . . . ' The real problem with news media's preoccupation with conflict, then, is not conflict *per se*, but the ritualistic manner in which it is staged as political spectacle, a meticulously scripted affair in which citizen-audiences have no role to play but bear witness to the fact that nothing ever seems to change.

Biggar explained that one of the most significant revelations from his experience heading up Leadnow as a political advocacy organization was the power of *unscripted opposition*, both as a means of generating public engagement with issues and having a meaningful impact upon the political process. He credited Ben Brandzel, founder and director of Online Progressive Engagement Networks (OPEN), for the idea: 'if you are going to actually change a decision, you have to create unscripted opposition that you did not expect. So it has to be of a different kind or a different intensity: that actually changes the political calculus. The same people doing the same thing does not.' Biggar identified four attributes of stories which are more likely to catalyze such opposition: first, 'a sense of bad dealing, that there is stuff going on behind closed doors'; second, 'a bit of mystery' and unpredictability, a sense in which outcomes are not predetermined; third, 'direct, tangible impact on real people and places that you care about'; and finally, 'a hero or heroic group that is taking on the folks who are engaged in bad dealing, and who will have a tangible impact on these very specific people in a place'. Based upon his own experience, Biggar argued that stories built out of these components were not only more likely to be circulated quickly and extensively through social networks, but also this distribution then generates 'political energy' which organizations such as Leadnow can help coordinate and focus in a politically effective manner (through campaigns, protests and other forms of direct action). 'It is hard for us to create energy. What is much easier is to create vehicles for the energy that already exists to be expressed and channelled as effectively as possible' (Biggar).

For Gillis, town hall meetings are particularly potent venues for building political efficacy as individuals come together as a community and realize – in some cases, for the first time – that others share their values, principles and, in many cases, their outrage. Describing his experiences as both a journalist and an occasional participant in such meetings, Gillis noted the political energy that is generated as people gather to express their anger about industrial and economic development which threatens the places and people they love. He described one such event which occurred in 2009 in Kaslo, a small community in the interior of the province, to discuss a proposal with government and industry representatives to construct a dam on a local river. While the project's purpose was to generate renewable electricity, it also would have had significant negative ecological impacts upon the local environment, destroying fish and wildlife habitat and spoiling a pristine watershed:

> Well, people were so fired up, and right away, people took the microphone and just took over the meeting. And they said, basically, this is how it is going to go. And one person after another . . . the First Nations said we are going to sue your asses off if this ever happens, and the other people were perhaps not quite as explicit as G****** from Cascade Falls. . . . 'You will never, ever build this project. Equipment will go missing. Shit will happen. We fucking guarantee it.' And you believed him, and it was 1,100 people, one after another. And guess what? That project never got built. And I filmed it and put it on [*The Common Sense Canadian* website, see Gillis 2010]. It is like a force multiplier, an echo effect . . . an amplification effect, a feedback loop.
>
> (Gillis)

Gillis spoke about reporting on dozens of similar town hall meetings to discuss a variety of projects, including plans to expand natural gas mining and construct massive LNG export terminals in coastal sites. Rafe Mair is a former B.C. politician, radio talk-show host and environmental activist, and a co-founder of *The Common Sense Canadian* with Gillis, and he is a frequent participant in these events. Gillis explained how Mair's interventions often follow a familiar script in which he fiercely challenges the legitimacy of shallow consultative processes designed to showcase a project's virtues rather than listen to local communities. Such outbursts, Gillis noted, 'gave people permission to be pissed off . . . because it told people that (a) they're not alone, there are 1,000 other people here who feel just like me; and (b) . . . you would see like an old grandmother or . . . maybe a meeker person who had never really spoken at the microphone, and they thought, "Yeah, what he said, that's how I feel," and then they would get up to the microphone, and all of a sudden they had permission to be like that.' Capturing and communicating this dynamic, visceral experience of empowerment is one of Gillis's primary objectives as an alternative media journalist. These moments

of confrontation serve as the performative accompaniment to 'cognitive liberation' in which citizens, both directly and indirectly, experience the transformative
pleasures of collective political agency in the face of structures, institutions and
processes which they had previously regarded as unassailable.

For Gillis, making political engagement meaningful and accessible involves
telling stories about those who are already engaged. Portraying political opposition to provincial plans to permit the construction of a network of dams was best
achieved, he argued, by putting 'a person to a river for every one of these [propos-
als]'. In one case, it was 'a bunch of these crazy kayakers who had been putting
their lives on the line to invoke the *Navigable Waters Protection Act*' to prevent
development of the river. It was 'through their eyes that I engage with their particular river. . . . I filmed these guys going like crazy and I talked to them why they
were doing what they were doing.' In other places, it was horse ranchers, local
farmers or even religious communities, and in each case, these individuals and
groups were not simply portrayed as victims, but as 'really interesting people . . .
living out their resistance' (Gillis).

The highly decentralized structure and practices of Leadnow means that the
group has to lean heavily upon the political labour and instincts of volunteers
to organize, coordinate and lead local campaigns across the country. Such campaigns often begin, Biggar explained, with:

> people signing a petition and then . . . 'Okay, there is going to be an action, and
> we are going to email you and we are going to ask if people can host events.'
> Okay, so now we have hosts. Now, we are going to email everybody else and
> ask them to go out to those events. Now we have a whole bunch of events.
> What you find is that it is surprisingly easy, and that it generates enormous
> numbers of first-time leaders. People who have never done anything before.

Underlying this organizational model is an elementary faith in the capacity and
desire of 'ordinary people' to get involved in issues they care about, a faith that has
been richly rewarded in the case of Leadnow. 'You have to trust these people. . . .
"OK, person who we have never met before, go lead a rally for us in front of an
MP's office." There is a lot of trust in that. And what we find is that that trust is
very rarely misplaced' (Biggar).

Beyond the effect of specific campaigns or events, they inevitably produce new
stories of political awakening and enjoyment which can, in turn, be fed back into
social networks, creating synergistic feedback loops of communication and activism which enhance perceptions of (collective) efficacy and hope. One of Leadnow's most powerful and effective messages, Biggar noted, took the form of a
testimonial from a campaign volunteer explaining how and why she had become
involved in the group and describing her experiences, including how they made
her feel. Stories of political participation (and associated experiences of empowerment and solidarity) – especially from 'ordinary' people with whom audiences
can empathize – are among the most powerful communications strategies for

countering the political cynicism that individuals cannot have any meaningful impact upon the political process:

> Showing people that others – and, preferably, others *like them*, with whom they can identify and empathize – are engaging in a particular form of action is a far better means of persuasion than simply explaining or asserting the need for or the benefits of that form of action. . . . According to this logic, the best means of increasing civic engagement would be to represent such behaviour as common, widespread, pleasurable and politically effective: in short, as *normal*.
>
> (Gunster 2012: 262)

Moral injunctions to 'get active' in climate politics are a common feature of environmental communication, and they may have some impact in terms of activating values and beliefs which can motivate people to get engaged. But they also risk amplifying feelings of guilt as people feel pressured to participate in activities which are unfamiliar to them or which they perceive will set them apart from their peers. Stories about people who already participate in climate politics – which not only describe why they are active but also how that experience makes them feel, has affected their identity, and changed how they understand and engage with the world – provide a much easier entry point into political engagement.

Kevin Washbrook, founder of the grassroots group Voters Taking Action on Climate Change (VTACC), noted that while he was initially somewhat reluctant to share his own personal story – 'I do not really enjoy talking about myself, I do not enjoy being part of the public realm' – he also discovered that grounding his activism in his own biography has enabled him to connect at a deeper level with audiences. 'People need to realize how easy this [civic engagement] is.' Sharing his experience as an 'ordinary guy' who started up an activist group with like-minded friends and neighbours is among his most empowering and hopeful messages. Accounts like these help bridge the gap between passive and active forms of citizenship, smoothing the transition from the former to the latter as people come to perceive different forms of democratic engagement as *normal* activities which those just like them are doing (and enjoying) in order to express and act upon their desire to do something about climate change.

While not all stories of civic engagement are necessarily rooted in conflict, direct confrontation of 'extractivism' – a phenomenon that Naomi Klein has dubbed 'Blockadia' (2014) – has become an accessible point of entry into climate politics for many, from local communities fighting specific projects to the rapidly growing divestment movement taking hold across university and college campuses. For Steven Schwarze, the rhetorical effects of melodrama – grounded in dramatic portrayals of intense social and political conflict – hold enormous potential to engage and mobilize audiences around environmental issues. He identifies five contributions as especially noteworthy (2006: 245–55). First, melodrama challenges the pervasive individualization of responsibility for environmental problems (Maniates 2001) by foregrounding the actions taken by large and powerful institutions: stories about innocent victims (human or otherwise) suffering as a consequence of corporate and/or government malfeasance can both clarify otherwise obscured relations of power and channel collective outrage towards specific targets. Second, polarization can disrupt existing forms of hegemony and thereby open up the potential for new political alliances to take shape around particular controversies: resistance to pipelines, for example, has significantly strengthened and deepened the ethical and political affinities between First Nations and the environmental movement in British Columbia. Third, melodrama can help shift public discourse around environmental issues on to a distinctively moral plane in which technical and economic discourses of risk often favoured by elites are displaced by elementary considerations of justice and fairness, right and wrong. Fourth, narratives of conflict often cultivate 'monopathy' – powerful emotions of identification, empathy and solidarity with the victims of environmental injustice and those who are fighting against it. Finally, melodrama can introduce and sustain complexity rather than simply reify and harden existing lines of conflict: 'it can expand the possibilities of controversy by bringing new issues to public attention, soliciting support from far-flung and previously inactive audiences, and complicating the grounds for public judgment' (Schwarze 2006: 254).

Solutions, conflict and the dialectic of engagement

These dichotomous views on the virtues and vices of conflict frames should not distract us from the underlying fact that both perspectives begin from the same diagnosis, namely, an impatience with commercial media's reliance upon conventional narrative stereotypes to simplify (and condemn) environmental politics as a Sisyphean battle between entrenched, professionalized coalitions of special interests. The routine use of conflict frames to filter reports about climate science, for example, has contributed to persistent and widespread misperceptions that scientists are deeply divided about the severity and even existence of anthropogenic global warming. The diminution of climate and energy politics to partisan bickering and stakeholder gridlock has also had toxic impacts upon levels of efficacy and hope among the general public (Hart and Feldman 2014). Such portrayals constitute a 'scorched earth' for civic engagement, ensuring that any optimistic messages around climate and energy policy that do actually make headlines are likely to fall upon barren ground. As P. Sol Hart and Lauren Feldman argue, 'even if news audiences . . . might come to believe that a proposed solution could ultimately work to help adapt to or mitigate climate change, the emphasis on political conflict in news coverage is likely to give the impression that the government will be unresponsive to calls for action and that the likelihood of implementing climate mitigation policies is very low' (2014: 314).

Differences arise, however, in how to respond to the narrow, one-dimensional, 'faux' politicization that currently drives much climate journalism. If reminding audiences about political conflict triggers scepticism about climate science, intensifies resistance to pro-environmental policies and behaviours, and exacerbates indifference, apathy and even disgust with political institutions, then perhaps the key signifiers, frames and storylines of such conflict should be avoided altogether (Lorenzen 2014). Instead, prioritizing shared values, common interests and concrete, pragmatic solutions which can attract bipartisan (or multi-partisan) support may not only enable progress around particular policies but also can more broadly kindle a (renewed) public faith in the capacity of political institutions to respond to social problems (Marshall 2014). This problem-solving disposition is likely to be especially resonant among enlightened and empowered 'influentials' who already possess the social, cultural and political capital to serve as agents of change.

And yet an equally compelling case can be made for revising and sharpening rather than repressing frames and stories of conflict. At one level, abandoning the terrain of political conflict demonstrates extraordinary naïveté about the possibility of building consensus around transformative social change, especially in countries with significant carbon extractive industries. More important, it sacrifices the conceptual, cultural and affective resources through which citizens, especially those who have been marginalized and excluded from traditional forms of politics, are motivated and mobilized to participate in collective action (Taylor 2000). Vigorous political debate, conflict and antagonism over competing visions of society are the very lifeblood of democratic politics (Mouffe 2005). Accordingly, championing technocratic, 'post-political' (Kenis and Lievens 2014) deliberations instead of attending to substantive political disagreements about how to address cascading ecological crises does little more than license the continuing hegemony of the capitalist (and consumerist) political economy that intensifies such crises (Foster *et al.* 2010; Klein 2014).

Investigating the communications strategies of U.S. environmental advocacy organizations, Luis Hestres identified a similar split over tactics which he traced to contrasting expectations and assumptions about social and political change. 'Organizations adhering to theories of change that emphasize subject-matter expertise and elite opinion persuasion are more likely to pursue online strategies that encourage low-threshold participation and facilitate the flow of ideas among elites' (2015: 207). Typical advocates of this approach include more traditional 'legacy' groups such as the Environmental Defense Fund (EDF) and the Natural Resources Defense Council (NRDC). Conversely, 'organizations embracing theories of change that emphasize broad-based, high-threshold participation are more likely to pursue online strategies that facilitate grassroots participation in the political process' (Hestres 2015: 207). Much more willing to embrace the mobilizing potential of conflict frames and narratives, 'Internet-mediated' groups such as 350.org and 1Sky have largely eschewed the path of building consensus within elite policy networks (and the general public as a whole) in favour of targeting and engaging a smaller community of individuals – the 'alarmed' issue public – who are already aware, deeply concerned and motivated to get involved in climate politics. These organizations 'shared a determination to help build a full-blown social movement to tackle climate change, anchored in a belief that the system was essentially non-responsive to conventional advocacy. They also shared a belief in the power of the Internet to help turn members of the climate issue public into committed, long-term activists' (Hestres 2014: 335). Indeed, the signature political achievement of 350.org was energizing opposition to the Keystone XL pipeline by catalyzing and leading a national campaign of civil disobedience.

Choosing between these two contrasting strategies (and associated theories of social change) can certainly pose challenges for environmental and political advocacy organizations, climate activists and even individuals weighing how to engage with their friends, neighbours, co-workers and social networks in conversations about climate and energy politics. And, as this chapter suggests, they are equally pressing for independent media organizations and journalists grappling with not only how to inform their audiences about climate and energy politics, but also to serve the needs and interests of those citizens keen to join with others in building a more just and sustainable future. Yet rather than champion either a 'facilitative' approach which emphasizes bipartisan solutions or a 'radical' role in which conflict frames mobilize political activism, the more important insight from this research may be the need to increase the presence and profile of *both* approaches in the ecosystem of journalism and advocacy communication. While there are obviously contradictions between these two strategies, this tension may well have a salutary impact upon levels of public engagement. At the very least, an emphasis upon either solutions or 'accessible' conflict constitutes a marked improvement over the cynical partisanship portrayed by commercial media.

Equally vital, however, is how the apparent tension between these frames can easily become a resonance in which the conceptual and affective force of one frame resounds to the benefit of the other. Expanding awareness of the practicality and benefits of particular solutions can strengthen the motivation to challenge those actors and institutions blocking their implementation. Likewise, mobilizing public interest and emotions around a specific controversy can stimulate greater attention to the role that potential solutions play in challenging (or affirming) the arguments of key players in a dispute. A stronger understanding of the technological, economic and policy dimensions of renewable energy can, for example, intensify willingness to participate in resistance to fossil fuel projects. Closing the 'hope gap', then, may ultimately depend less on prioritizing a particular frame or communications strategy and more on an expanding independent media ecosystem which can amplify a diversity of alternatives to the political cynicism of commercial media.

Notes

1 Research for this chapter was undertaken as part of a larger project exploring the impact of news about climate politics upon the public which was undertaken as part of the Climate Justice Project, led by the Canadian Centre for Policy Alternatives (CCPA) and the University of British Columbia (UBC) and funded by the Social Sciences and Humanities Research Council (SSHRC). Additional financial assistance was provided by the Dean of Graduate Studies at Simon Fraser University (SFU). The author is grateful for the contributions and advice of the team lead of the larger project, Kathleen Cross, as well as Robert Hackett, Shannon Daub and Helena Krobath. The interviews for this project were undertaken by the author, Robert Hackett and Kathleen Cross; however, responsibility for analyzing the results lies with the author alone. A preliminary version of this research was presented at the 2015 International Environmental Communication Association conference in Boulder, Colorado.
2 *The Tyee* receives between 800,000 and 1 million page views and between 200,000 and 400,000 unique visitors per month (*Tyee* 2016) which compares favourably to an average paid daily circulation in 2014 of just over 86,000 for *The Vancouver Sun*, the region's leading commercial broadsheet newspaper (Newspapers Canada 2014).

References

Beers, D. (2010) 'Andrew Nikiforuk is Tyee's first writer in residence', *The Tyee* (July 15). Accessed at http://thetyee.ca/News/2010/07/15/Nikiforuk.

Berglez, P. (2011) 'Inside, outside, and beyond media logic: Journalistic creativity in climate reporting', *Media, Culture & Society* 33(3): 449–65.

Chen, S. and Gunster, S. (2016) '"Ethereal carbon": Legitimizing liquefied natural gas in British Columbia', *Environmental Communication* 10(3): 305–21.

Cross, K., Gunster, S., Piotrowski, M. and Daub, S. (2015) *News Media and Climate Politics: Civic Engagement and Political Efficacy in a Climate of Reluctant Cynicism*, Vancouver, BC: Canadian Centre for Policy Alternatives.

Curry, A. and Hammonds, K. (2014) 'The power of solutions journalism', *Engaging News Project/Annette Strauss Institute for Civic Life at the University of Texas Austin*. Accessed at http://engagingnewsproject.org/enp_prod/wp-content/uploads/2014/06/ENP_SJN-report.pdf.

Curry, A., Jomini, N., Stroud, N. and McGregor, S. (2016) 'Solutions journalism and news engagement', *Engaging News Project/Annette Strauss Institute for Civic Life at the University of Texas Austin*. Accessed at http://engagingnewsproject.org/enp_prod/wp-content/uploads/2016/03/ENP-Solutions-Journalism-News-Engagement.pdf.

Downing, J.D.H. (2001) *Radical Media: Rebellious Communication and Social Movements*, Thousand Oaks, CA: Sage.

Edge, M. (2007) *Asper Nation*, Vancouver, BC: New Star Books.

Fast, T. (2014) 'Stapled to the front door: Neoliberal extractivism in Canada', *Studies in Political Economy* 94(1): 31–60.

Forde, S. (2011) *Challenging the News: The Journalism of Alternative and Community Media*, New York: Palgrave Macmillan.

Foster, J.B., Clark, B. and York, R. (2010) *The Ecological Rift: Capitalism's War on the Earth*, New York: Monthly Review Press.

Gillis, D. (2010) 'Glacier-Howser contract canceled!', *The Common Sense Canadian* (November 15). Accessed at http://commonsensecanadian.ca/glacier-howser-contract-canceled/.

Gunster, S. (2011) 'Covering Copenhagen: Climate politics in B.C. media', *Canadian Journal of Communication* 36(3): 477–502.

—— (2012) 'Visions of climate politics in alternative media', in A. Carvalho and T.R. Peterson (eds), *Climate Change Politics: Communication and Public Engagement*, Amherst, NY: Cambria Press, pp. 247–77.

Hart, P.S. and Feldman, L. (2014) 'Threat without efficacy? Climate change on U.S. network news', *Science Communication* 36(3): 325–51.

Hestres, L. (2014) 'Preaching to the choir: Internet-mediated advocacy, issue public mobilization, and climate change', *New Media & Society* 16(2): 323–39.

—— (2015) 'Climate change advocacy online: Theories of change, target audiences and online strategy', *Environmental Politics* 24(2): 193–211.

Kenis, A. and Lievens, M. (2014) 'Searching for "the political" in environmental politics', *Environmental Politics* 23(4): 531–48.

Klein, N. (2014) *This Changes Everything: Capitalism vs the Climate*, Toronto: Knopf.

Lorenzen, J. (2014) 'Convincing people to go green: Managing strategic action by minimizing political talk', *Environmental Politics* 23(3): 454–72.

Maniates, M. (2001) 'Individualization: Plant a tree, buy a bike, save the world?', *Global Environmental Politics* 1(3): 31–52.

Marshall, G. (2014) *Don't Even Think about It: Why Our Brains Are Wired to Ignore Climate Change*, New York: Bloomsbury.

McCreary, T. and Milligan, R. (2014) 'Pipelines, permits and protests: Carrier Sekani encounters with the Enbridge Northern Gateway Project', *Cultural Geographies* 21(1): 115–29.

Mouffe, C. (2005) *On the Political*, New York: Routledge.

Newspapers Canada (2014) 'Circulation report: Daily newspapers 2014'. Accessed at http://www.newspaperscanada.ca/sites/default/files/2014_Circulation_Report-Daily_ Newspapers_in_Canada_FINAL_20150603_0.pdf.

Nikiforuk, A. (2012) *The Energy of Slaves: Oil and the New Servitude*, Vancouver, BC: Greystone Books.

Olausson, U. (2011) '"We're the ones to blame": Citizens representations of climate change and the role of the media', *Environmental Communication* 5(3): 281–99.

Olausson, U. and Berglez, P. (2014) 'Media and climate change: Four long-standing research challenges revisited', *Environmental Communication* 8(2): 249–65.

O'Neill, S., Williams, H.T.P., Kurz, T., Wiersma, B. and Boykoff, M. (2015) 'Dominant frames in legacy and social media coverage of the IPCC Fifth Assessment Report', *Nature Climate Change* 5: 380–5.

Rhodes, E. and Jaccard, M. (2013) 'A tale of two climate policies: Political economy of British Columbia's carbon tax and clean electricity standard', *Canadian Public Policy* 39(2): S37–S51.

Rosenstiel, T. (2015) 'Reporting the whole story: Nine good questions with David Bornstein of SJN', *American Press Institute*. Accessed at https://www.americanpressinstitute. org/publications/good-questions/moving-toward-whole-story-9-good-questions-david-bornstein-solutions-journalism-network/.

Salazar, D. and Alper, D. (2002) 'Reconciling environmentalism and the left: Perspectives on democracy and social justice in British Columbia's environmental movement', *Canadian Journal of Political Science* 35(3) (September): 527–66.

Schäfer, M.S. and Schlichting, I. (2014) 'Media representations of climate change: A meta-analysis of the research field', *Environmental Communication* 8(2): 142–60.

Schwarze, S. (2006) 'Environmental melodrama', *Quarterly Journal of Speech* 92(3) (August): 239–61.

Smith, A. and MacKinnon, J. (2007) *The 100-Mile Diet: A Year of Local Eating*, Toronto: Random House Canada.

Smith, J. (2005) 'Dangerous news: Media decision making about climate change risk', *Risk Analysis* 25(6): 1471–82.

Taylor, D. (2000) 'The rise of the environmental justice paradigm: Injustice framing and the social construction of environmental discourses', *American Behavioral Scientist* 43(4) (January): 508–80.

The Tyee (2016) 'Tyee ad kit'. Accessed at http://www.thetyee.ca/Ads/2011/09/28/Tyee_AdKit.pdf.

Tyee Solutions Society (2015) Accessed at http://www.tyeesolutions.org/.

Upton, J. (2015) 'Media contributing to "hope gap" on climate change', *Climate Central*, (March 28). Accessed at http://www.climatecentral.org/news/media-hope-gap-on-climate-change-18822.

Interview participants

Given the limits of space, it was not possible to include comments from all interviews in the text of this chapter. However, the arguments of the chapter were informed by an analysis and consideration of the comments from all participants.

Alternative media

David Beers, Founding Editor, *The Tyee*
Geoff Dembicki, Lead Sustainability Writer, *The Tyee*
Damien Gillis, Editor, *The Common Sense Canadian* and documentary filmmaker
Chris Wood, Coordinating Editor, *Tyee Solutions Society*
Linda Solomon Wood, Founder, Editor and Publisher, *Vancouver Observer*

Environmental advocacy NGOs

P.J. Partington, Climate Policy Analyst, *Pembina Institute* (Toronto)
Kevin Sauvé, Communications Lead, *Pembina Institute*, B.C.
Ben West, Executive Director, *Tanker Free B.C.*
* Communications Specialist, *B.C. Environmental advocacy NGO*

Political advocacy NGOs

Jamie Biggar, Executive Director, *Leadnow*
Alan Dutton, Spokesperson, *Burnaby Residents Opposing Kinder Morgan Expansion* (BROKE)
Kevin Washbrook, Founder, *Voters Taking Action on Climate Change* (VTACC)

* All participants were given the opportunity to participate anonymously in this research. One participant chose this option, and this individual has been identified as a communications specialist with a B.C.-based environmental advocacy organization.

Chapter 6

Australian independent news media and climate change reporting

The case of COP21

Kerrie Foxwell-Norton

> *Security, defiance, hope, impotence. Paris demonstrated a whorl of contention and affect. As #COP21 progresses, climate activists will find a way to express the voice of civil society. For there can be no peace on earth while our atmosphere is choked with carbon.*
>
> (Liz Conor, *New Matilda* Columnist, December 3, 2015, Paris)

Climate change is a wicked problem. Debate about the issue creates a complex terrain, with different meanings emerging that muddy our understanding of the science, the impact of climate change and the possible (and appropriate) responses (Hulme 2009, 2015). Regardless of location, research into climate change communication via journalism, media or otherwise is a lesson in the sobering humility of the agonism and plurality of public sphere debate (Mouffe 2013). The oft-cited 'scientific consensus' that climate change is anthropogenic is the pinnacle of clarity, whence complexity, context and contingency abound. In other words, what to do and who needs to do it and how and why this should be done complicate what we might mean by 'action'. Hulme (2015: 900) explains:

> The goals of 'action' on climate change might therefore be, inter alia, to limit global warming to two degrees, to deliver creation care, to design a planetary thermostat, to transform civilization or to safeguard economic growth – or indeed to secure fair growth, zero growth or de-growth. All of these goals have prima facie credibility since they emerge from different readings of what climate change is about, inspired by different cosmologies and ethical or political values. They emerge from different judgments being passed on the facts. Far from there being the possibility of a singular 'decisive political action' on climate change, the strategic goals of policy interventions are inevitably multivariate because they are shaped by different worldviews and different narratives of good human living.

The complexity of climate change should not be underestimated or indeed simplified and smoothed away. Managing the antagonisms that characterize a vibrant healthy communicative democracy (Hackett 2000) where the goal is citizen participation in decision-making is a global challenge with local ramifications for climate change communication. Communicating this pluralism and empowering the diverse expressions of communities and society more broadly has long been the remit of community and alternative media. What, then, are the realized or potential roles for the journalisms of alternative and independent media in communicating and/or instigating urgent climate change action? Can these media bring hope and security to civil society as *New Matilda* columnist Liz Conor outlined previously? Calling citizens to participate, 'to do something' (Forde 2011: 165), to empower citizens in Western liberal democracies with an opportunity to be heard and, in turn, to hear a diversity of ideas is a recurring theme in the alternative and community media literature (see also Atton 2002; Carpentier *et al.* 2003; Couldry and Curran 2003; Fuchs and Sandoval 2015; Rodriguez 2001).

In this chapter, I explore Australian independent and alternative news media coverage of the 2015 United Nations Climate Change Conference, commonly called COP21 or CMP11, that was held in Paris, France, from November 30 to December 12, 2015. In contrast with previous studies that focused on mainstream news media coverage of climate change in Australia, this chapter interrogates Australia's independent media sector at a critical moment in international climate change policy and action. News articles were collected from online independent news media sites in Australia: *New Matilda*, *Crikey*, *Green Left Weekly* and *Independent Australia*. The chapter begins with a brief introduction to climate change in Australia, in recognition that local contexts affect the content and character of climate change discussion. I then turn to the idea of community, independent, alternative media and what we might expect in an era of climate crisis. Salient characteristics of journalist and publication intent, sourcing practices, political engagement, critique and citizen action are identified as *sine qua non* to these media and are established as the guiding criteria for this investigation. Australia has one of the most concentrated media ownership environments in the world, dominated by Rupert Murdoch's News Ltd. Diversity from alternative and independent media in Australia thus makes a particularly interesting case study. While there is increasing research on Australian mainstream media and climate change (e.g. Bacon 2013; Chubb 2012, among others), independent and alternative media outlets are yet to receive the same attention. This chapter begins to fill this gap in our understanding of climate change coverage and the Australian news media.

This investigation finds that Australian independent media exhibit a clear motivation to incite citizen action and critique of the hegemonic power structures that have catapulted humanity toward environmental catastrophe. In Australia, this critique is directed at the collusion between governments and a once-booming fossil fuel industry, coupled with the elevation of climate change consequences for Australia's much loved 'natural wonders' and climate-vulnerable South Pacific neighbours. However, I also argue that, like mainstream media, independent media tend to communicate climate change as an 'imminent disaster' which, as others have asserted (Lester and McGaurr 2013; Painter 2013), is disempowering for audiences, and journalists may be able to do better. Opportunities thus exist to extend critique and citizen action to 'visioning alternative futures' and 'fostering authentic hope' (Moser 2015: 407). These are climate change messages that make action possible and – based on this research – Australia's independent media are the most likely to lead and deliver.

Climate change communication: News media in the Australian context

Australia is a peculiar participant in climate change negotiations and policy. Given the performance of Australian governments in the international arena, notably the decade-late entry into the Kyoto Protocol in 2007, statements of climate change scepticism from our nation's conservative political leaders, and the ongoing approval of gigantic coal mining projects, observers could be forgiven for assuming that Australia is a nation of carboniferous climate sceptics. Somewhat surprisingly and despite consistent research that has highlighted the dominant presence of climate scepticism in Australian mainstream news media (Bacon 2011, 2013; Chubb 2012; Lester and McGaurr 2009, 2013; McGaurr et al. 2013; McKnight 2010), most Australians believe in climate change. But that is about where the good news ends. In 2015, Australia's chief scientific research organization, the CSIRO, published its longitudinal study (Leviston et al. 2015) finding that nearly, 80 percent of Australians, relatively consistent across gender and geography, believe that climate change is happening. Around 45 percent believe that climate change is caused by human impacts. However, nearly 40 percent believe in climate change but think it is a natural fluctuation that is unrelated to human activity or impacts. The remaining small group – about 15 percent – thought that climate change was not happening at all or did not know. In more bad news, Tranter and Booth's cross-national study (2015) found that Australia had the highest proportion of climate sceptics followed by Norway, New Zealand and the U.S. For a nation that has prided itself on both its natural 'resources wealth' and beauty, climate change is a jagged pill to swallow. Australians are the highest emitters per capita of carbon in the world, in part a reflection of our fossil fuel resources industry (Garnett 2015). Australia's mostly remote mining operations in the 'dead heart' of the country where population is sparse contrasts with the lush, livable and loved coastal landscape. As the world's largest island continent with 37,500 kilometres of coastline, more than 80 percent of the population lives within 50 kilometres of the coast (Australian Bureau of Statistics 2012) and most of these on the east coast. Coastal nations are particularly vulnerable to climate change and forecast sea level rise. For a coastal nation such as Australia, climate change planning and policy are particularly urgent. Indeed, climate change impacts have arrived for other coastal nations in the region, namely Kiribati, Marshall Islands and Tuvalu and the many Small Island Developing States. In Australia, the irony of a national imagination that regularly boasts its natural wonders while simultaneously pillaging its natural fossil fuel resources for economic benefit is difficult to communicate – whatever the journalism, whatever the media. Understanding this schism in the Australian national psyche provides some critical context to climate change reporting.

The clarity of climate change communication more broadly has been further hampered by the impact of business interests and partisan politics on debate and discussions. Climate change in Australia confronts an established and, until recently, booming mining industry tied to international markets and corporations, especially evident in China. This status quo was first challenged when, in 2007, incoming Labor Prime Minister Kevin Rudd ratified the Kyoto Protocol as his government's first formal action. To combat climate change, the Rudd government in 2008 proposed an emissions trading scheme – the Carbon Pollution Reduction Scheme (CPRS) – based on the recommendations of the Garnaut Climate Change Review (2008). The CPRS marked the beginning of a particularly chaotic period in Australian politics with extraordinary internal instability within the Labor Party. In 2010, Julia Gillard became Australia's first female Prime Minister, replacing Kevin Rudd in a midnight coup that still attracts controversy (see Walsh 2013). Gillard introduced a climate change policy that entailed a carbon tax, a policy she had excluded before her election. While not within the scope of this chapter, the response from the political opposition, corporate mainstream news media and vested mining/business interests to both the Rudd and Gillard governments' climate policies was brutal and unrelenting (see Chubb 2014). In September 2013, and following further Labor Party internal instability (before the election Rudd was re-elected Labor leader, thus ousting Gillard) the conservative Liberal Party–National Party coalition government under the leadership of Tony Abbott was elected on a promise to abolish that policy. The Abbott government dismantled the climate change policy proposals and the Climate Change Commission that had been established by the Rudd/Gillard governments to provide reliable and authoritative advice on climate science and policy. Led by eminent Australian scientist and global warming activist Professor Tim Flannery, the Climate Change Commission was re-launched as the (unfunded) Climate Council within a week of its dismantling. Abbott introduced a much watered-down Emissions Reduction Fund and Direct Action Plan that is an 'opt-in' plan and does not carry penalties for emissions.

Mainstream media coverage during this tumultuous period in Australian climate change policy and debate has received some academic attention. Wendy Bacon surveyed 10 mainstream Australian newspapers and found clear evidence of bias towards climate scepticism and a preponderance of political and business sources, particularly those related to fossil fuel–dependent industries. Her 2011 report found that business sources trumped all other Australian civil society sources, including unions, NGOs, think tanks, activists, members of the public, religious spokespeople, scientists and academics (Bacon 2011: 14). Similarly, in 2013, Bacon found poor reporting practice of climate science and that climate scepticism continued to get 'substantial favourable exposure in Australian mainstream media reporting' (Bacon 2013: 15). Providing the Australian data for James Painter's (2013) six-nation study, McGaurr and Lester delivered a nuanced analysis of Australian newspaper coverage of the first two reports of the Intergovernmental Panel on Climate Change (IPCC) in 2007, IPCC's SREX Report in 2012 and the reporting of Arctic sea ice decline in the mainstream corporate press. Their findings concur with Bacon (2011, 2013), though adding a dimension in providing targeted analysis of the reporting of risk and uncertainty – and practical strategies for reporting:

> [I]nstead of tempering themes around possible disaster with information about opportunities, the Australian press has often given space to sceptics, or reported uncertainty without providing sufficient context for that uncertainty. Australia had the most articles, and the highest percentage of articles, with sceptics represented. Along with reports from India, Australian articles were also least likely to temper uncertainty frames with references to increasing scientific certainty.
>
> (McGaurr *et al.* 2013: 26)

Myra Gurney (2013) has provided a critical discourse analysis of the moral and ethical dimensions of climate change debate in key speeches of Australian political actors in the particularly tumultuous years of 2007 through 2011. She found that the moral dimension of climate change action was trumped by the more strategically powerful economic frame. Philip Chubb's comparative study (2012) of Australian mainstream media[1] coverage of the 2009 COP16 Copenhagen and 2011 COP17 Durban conferences found domestic political voices commandeered reporting. Interestingly, he found a dramatic decrease in coverage between these two periods that at first glance may indicate that climate change had disappeared from Australian debate. On the contrary, 2011 marked a period of intense and bitter debate on Australian climate policy, with the Federal Labor government proposing one of the world's first carbon tax policies (Chubb 2012: 180). Chubb concludes that 'those powerful elements of the business community whose leaders battled the carbon tax so bitterly fought side by side with the anti-science forces in the nation's political leadership and media' (p. 193; see also Chubb and Bacon 2010). Australian research follows international research on climate change and the media, with a focus on mainstream news media. While there are a few exceptions internationally (see e.g. Gunster 2011, 2012; Kenix 2008; Smeltzer 2008 and in this work; Foxwell-Norton 2015; Hopke 2012), the descriptive and normative role of community, alternative, citizen's, and radical media in climate change debates is an area ripe for further research (Foxwell-Norton 2016). As will be evident in the following data, the contrast between alternative media and their mainstream contemporaries is stark in the Australian context.

Defining independent media: Expectations for climate change communication

Subtleties of definition are perennial in the field of alternative and community media (Atton 2002; Downing 2001; Howley 2005; Rodriguez 2001). Overall, the extraordinary heterogeneity of such journalism and media and their creativity, innovation and resourcefulness are difficult to quantify, let alone qualify with a steady definition. From the relatively privileged examples which emerge from politically stable developed Western nations to war-torn conflict and trauma where community media literally save lives (Rodriguez 2011), the landscape is vast and diverse, expressing the multiplicity of human experience. In recent decades, these definitions are further complicated by the emergence of new social media and their potential to create independent avenues free of any connection to legacy (i.e. pre-digital) media. This has allowed avenues of communication for citizens, organizations and social and environmental movements that do not rely on traditional media hierarchies and journalism practices (Hutchins and Lester 2015). Legacy news media, however, still perform a critical role in the communication of environmental issues and specifically climate crisis (see Lester 2015: 239) as part of this 21st century new media ecology that provides mammoth possibilities for the communication and dissemination of environmental issues and protest (Cottle 2008: 855). For this chapter and following Rodriguez (2001), I focus on alternative and independent media as entities unto themselves, investigating their practices without the impulse to consider 'what they are not' that often accompanies comparisons to other media. Still, some comparison is unavoidable in the interests of acknowledging existing research, but this is done primarily to highlight the potential shape of journalisms for climate crisis.

Discussing the 'names and naming' of what is variously termed citizen's, independent, grassroots, community, radical and/or alternative media, Atton (2015: 3) draws on the idea of a 'bounded discourse' (Couldry 2010: 6) to highlight that any term is limited and changes within the cultural practices that give it meaning for people across time and places. Situatedness, position and cultural context – historical, political, economic, social and the oft-assumed scientific – influence how we would define and distinguish the news media we investigate and their reporting of environmental issues (see Lester 2010). This idea applies to the realm of alternative media reporting of climate change also and, here, how I define and distinguish alternative, independent media from 'dependent' or mainstream media. Elasticity in definitions aside, some key themes emerge, and they relate directly to the potential of these media to promulgate calls for climate change action. Relatively consistent across the literature is the role of these media in encouraging, empowering and inviting citizens to participate in public sphere debate and, further, to take some form of action. This action stems from the communication of critique and dissatisfaction with the existing power structures. From this basis of dissatisfaction, possibilities for action can be explored and citizens empowered to seek alternatives (either personally and/or politically) and plans and/or solutions to climate change crises enacted.

Such empowerment occurs via two broad and recurrent observations of these media: first, in *production*, through the ownership and/or broadcast of citizen's own media. Here, the distance between media producer and media consumer is dissolved or reduced, empowering citizens with the opportunity to broadcast their ideas and opinions (Forde *et al.* 2009; Fuchs and Sandoval 2015: 167). These ideas are often marginal and/or disempowered by mainstream corporate media reporting because of their bottom-line pursuit of profits, advertising and increased market shares. Alternative and independent media are relatively free of the bound- aries imposed by commercial imperatives and this allows these media to func- tion differently in relation to the dominant hegemonic discourses associated with neoliberalism and industrial capitalism (Atton 2002; Downing 2001; Rodriguez 2001, 2011). Second, empowerment occurs in *reception*, by providing a space for the broadcast of a multiplicity of voices, exposing audiences to ideas that expand the public discourse, inviting different interpretations of ideas, events and phe- nomena. In the production and reception of a diversity of media texts, alternative media provide an ongoing challenge and check on the dominance of hegemonic forms, be they related to capitalism and neoliberalism or, more specifically, to race, gender, sexuality and ethnicity. Here it is progressive media (Downing 2001) that operate for and within civil society, fostering the diversity of ideas critical to a healthy functioning and radical democracy that is enacted in the everyday actions of citizens (Laclau and Mouffe 1985). As such, these media correspond to Chris- tians *et al.*'s (2009) facilitative and radical roles suggested throughout this book as those most capable of inciting public action on climate crisis.

To reiterate briefly, the facilitative role sees journalism promote dialogue and active citizenship, not only reporting society as a distant detached observer (characteristic of the monitorial role) but rather actively trying to improve 'civil society actions and associations' (2009: 158). This role accepts the 'multidimen- sional character of community' and journalism's role in the (re)production of culture(s), and so it is mindful of pluralism and ensures this is reported and repre- sented by the news media. Conversely, the radical role rationale is:

> to expose to public opinion the concentration of social power, especially regarding the democratic procedures of collective decision-making. This implies a persuasive dimension, with attempts to mobilise public opinion and public action toward the redistribution of social power . . . 'radical' refers here to a perspective that literally goes to the roots of the power relations in society, challenging the hegemony of those in power and offering an alternative vision not just for some building blocks but for the whole structure of society.
>
> (Christians *et al.* 2009: 181)

The facilitative and radical roles focus on civil society operations and citizen empowerment and are acknowledged by Christians *et al.* (2009) as containing some overlap. They are differentiated by their purpose, where the facilitative role promotes citizen dialogue while the radical role seeks to mobilize opinion against societal power structures (p. 190). Usefully, they distinguish radical journalism from radical media, arguing that while radical media are rare, their tradition of journalism intent has a longer history. In a similar vein, Forde (2011, 2015: 293) proposes alternative journalism be recognized as an 'enduring form' with a 'political act' at its core. Atton and Hamilton (2008: 1) also identify a distinctive 'alternative journalism' explaining that it 'proceeds from dissatisfaction not only with mainstream media coverage of certain issues and topics, but also with the epistemology of news'. This discontent has led to alternative practices to mainstream journalism and media that reject conventions of news sources and representations and accepted 'professional' norms, including objectivity, elitism, commercial objectives and the passive audience. Forde (2011) extends those characteristics of alternative journalism to include citizen action and activism, with journalists often participating as activists themselves, supporting and contributing to progressive social and environmental movements. These characteristics are evident in the four alternative and independent media mastheads examined in this chapter – *New Matilda*, *Crikey*, *Independent Australia* and *Green Left Weekly*. To illustrate, I have looked to their website descriptions for rationales of their content and purpose. Their descriptions are instructive:

New Matilda (NM)

We believe that robust media are fundamental to a healthy democracy – and there's never been a more important time for independent media in Australia. With shrinking media diversity, and huge changes underway in delivery, there are fewer and fewer outlets publishing independent-minded journalism like ours. . . . Our readers are our investors. *NM* is predominantly reader-funded and remains fiercely independent, with no affiliation to any political party, lobby group or other media organisation.

(https://newmatilda.com/about-us/)

Crikey

Crikey's aim is very simple: to bring its readers the inside word on what's really going on in politics, government, media, business, the arts, sport and other aspects of public life in Australia. *Crikey* reveals how the powerful operate behind the scenes, and it tackles the stories insiders are talking about but other media can't or won't cover. *Crikey* sees its role as part of the so-called fourth estate that acts as a vital check and balance on the activities of government, the political system and the judiciary. In addition, *Crikey* believes the performance and activities of business, the media, PR and other important sectors are worthy of public scrutiny.

(https://www.crikey.com.au/about-crikey/)

Independent Australia (IA)

IA is a progressive journal focusing on politics, democracy, the environment, Australian history and Australian identity. It contains news and opinion from Australia and around the world. *IA* . . . believes Australians are short-changed by the mass media – and so it dedicates itself to seeking out the truth and informing the public. . . . *IA* is also opposed to partisan politics and supports Independent politicians.

(https://independentaustralia.net/about)

Green Left Weekly (GLW)

In these days of growing media concentration, *GLW* is a proudly independent voice committed to human and civil rights, global peace and environmental sustainability, democracy and equality. By printing the news and ideas the mainstream media won't, *GLW* exposes the lies and distortions of the power brokers and helps us to better understand the world around us. . . . Most importantly, *GLW* is a campaigning paper: it helps strengthen the anti-racist, feminist, student, trade union, environment, gay and lesbian, civil liberties and anti-imperialist movements by linking the issues and activists, and by letting people know how they can join others in action for change.

(https://www.greenleft.org.au/about)

Evident is the explicit commitment to the facilitative and radical roles through a focus on the Fourth Estate as a check on the established order which, at the current juncture, must include neoliberalism as the prevailing political and economic orthodoxy. While the publications vary in their degree of 'radical', from the more militant *GLW* to the far more moderate *Crikey*, their intention is to report independently about, and independent of, Australia's mainstream news media. Their commitment to civil society organizations and democracy, to challenging hegemonic structures and to providing news free of commercial pressures has been the guiding principle in this investigation of Australian independent media coverage of COP21.

Methodology

In total, 45 articles were collected over a four-week period (23 November – 18 December 2015), using specific key terms, from the four independent online news outlets. Their varying frequency of publishing affected the number of articles collected from each. The articles comprised the following: 22 from the tri-weekly digest *New Matilda* which publishes every Tuesday, Thursday and Sunday; 9 from the daily news site *Crikey*; 10 from the daily *Independent Australia*; and 4 from the weekly *Green Left Weekly*.

Several databases were used to cross-check article retrieval, including Factiva, Proquest and Informit, supplemented by searches within individual publications and broader Google searches. Search terms were 'COP21', 'Paris' and 'Climate Change'. Both news and comment pieces (editorial and opinion) were collected.[2] Comment pieces were included following Bacon's (2011, 2013) *Sceptical Climate* studies that noted the strong influence of 'comment' (columnists in particular) in promoting climate scepticism in Australia. The COP conferences are a repeated global media event and, as such, offer an opportunity to explore journalism's role in the sphere of global climate politics. Given their annual meeting, they also provide a reliable platform to analyze and compare journalism and media practices, both within and across media platforms and geographies and over time (see Eide and Kunelius 2012; Eide *et al*. 2010).

Given the volume of articles available for analysis – representing the corpus of material produced about COP21 during that period – a form of qualitative content analysis was used to identify trends in the data. Qualitative content analysis 'is defined as a research method for the subjective interpretation of the content of text data through the systematic classification process of coding and identifying themes or patterns' (Hsieh and Shannon 2005: 1278; Mayring 2000; Schreier 2014). More specifically, a directed qualitative content analysis was adopted here which draws on existing theory and research to identify codes and categories (Hsieh and Shannon 2005). A coding schedule was developed based on the review of previous Australian media and climate change research and the distinguishing and purported characteristics of independent and alternative media. Sources were the principal means used to code all articles covering broad and specific categories related to 'politicians', 'industry', 'environmental movement organizations', 'United Nations', 'think tanks', 'scientists' and other smaller categories. These categories are an amalgamation of previous coding categories established by Bacon (2011, 2013), Chubb (2012) and Lester and McGaurr (2013). To these categories, 'citizen action' and 'critique' were added as distinctive characteristics of independent news media established in the alternative and community media literature. News sources are the building blocks of news structure, and 'no analysis of news media content is complete without a close look at the sources of that content' (Reese 1994: 87). Focusing on sources emphasizes that much of the actual 'power' of media comes from their ability to 'amplify the views of certain powerful sources' (Reese 1994: 87).

A research assistant collected and coded all articles with three distinct phases of coder reliability conducted alongside ongoing discussion and communication characteristic of content analysis. A coding schedule elicited some limited quantitative data in terms of the use of sources and frequency of certain perspectives, and this is presented first to provide some basis to the findings. The content analysis – albeit of a small corpus of material that may be expected of smaller independent publications with less frequent publishing schedules – was accompanied by qualitative textual analysis and, in particular, critical discourse analysis. This approach links the use of language to particular social, cultural and political contexts. It seeks an explanation for 'the links between texts and social relations, distribution of power, and dominant values and ideas' (Carvalho 2008; Carvalho and Burgess 2005: 1461; van Dijk 2000). Arguably, the most important aspect of understanding media coverage is understanding the 'interactions between different institutional arenas and how these influence the discourses and representations of environmental news' (Cottle 2008: 77). In a similar vein, Bacon and Nash (2012: 245) assert the value of seeing the media in relation to other 'fields of social power'. In previously providing some cultural and political context to climate change in Australia, these 'relations of definitions' are explored and provide a basis to investigate these media and their interactions with other 'institutional arenas' and to consider the consequences for reporting practices.

Broadly, the key themes explored in this study are the distinctive motivations and intentions of the journalists and their publications, sourcing practices which prioritize groups marginal to COP debates proper and mainstream media coverage, and the coverage of citizen action and explicit critique of governments and the fossil fuel industries that contribute disproportionately to climate change. These themes are explored further in subsequent discussions.

Journalists, commentators and intent

Both journalists and commentators shared a commitment to Fourth Estate principles and were often combining activism with their writings. Their current profiles and histories indicate an approach to journalism that correlates with the facilitative and radical roles to empower citizen dialogue and action and to challenge hegemonic structures, particularly from those marginalized or disadvantaged by the prevailing social, political and/or economic orthodoxy. To illustrate, creator and managing editor of *IA* David Donovan's citizen activism is evident in his role as former media director and vice-chair of the Australian Republican Movement. *IA*'s deputy editor and environment editor, Sandi Keane, has 'been involved in the environmental movement since the 1970s, and in 2009, she was invited as one of Victoria's environment community leaders to an invitation-only breakfast with Al Gore. Her blog can be found at http://scepticslayer.wordpress.com with the title 'Carbon Claptrap . . . and other bunkum . . . slaying the sceptics and debunking the deceivers'. Alongside her position at *IA*, Keane contributes to other publications in this sample, *Crikey* and *NM*. Also working across publications including *Crikey* and others, Wendy Bacon is contributing editor to *NM* and is also the academic author of the *Sceptical Climate* reports cited here. On her website (see http://www.wendybacon.com), Bacon explains her role as:

> an investigative journalist who is also a political activist. This means that I want my journalism to be useful to those who resist abuses of power and seek social justice rather than supporting existing power structures, which is

what most journalism does. My emphasis is on information that I hope will empower people to take action.

Chris Graham, owner and editor of *New Matilda*, was the former founding editor of the *National Indigenous Times* and *Tracker*. Both of these publications represented a critical facet of Australia's Indigenous media sector, bringing First Nations' voices, perspective and knowledge to the Australian mediascape. Reflecting on his two years in the *NM* role in 2016, Graham describes the core values of this independent publication, including its attitude to climate change: 'we're passionate about climate change because if we don't stop destroying the planet then life as we know it ends. All we'll end up passing on to our children is social and economic ruin.' Graham (2016) distinguishes his publication because it does not 'pretend that [it's] providing balance by publishing people with extremist views, like Paul Sheehan, or Andrew Bolt, or Miranda Devine, or Gerard Henderson or any number of right-wing conspiracy theorists.'

Crikey commentator Erwin Jackson, who produced three comment pieces during the sample period, is the Deputy CEO of Australia's Climate Institute, whose professional experience is in climate change policy and programs to reduce greenhouse gas emissions. Similarly, *IA* published a comment piece from Giles Parkinson, founder and editor of RenewEconomy.com.au, a website focusing on clean energy news and analysis, as well as climate policy. Both *GLW* and *IA* produced their own articles but also aggregated news from other sites, including The Climate News Network (see http://climatenewsnetwork.net) that publishes a daily news story on climate and energy and also provides journalist training and advice on reporting climate change. Another comment piece from *IA* (December 1, 2015), was authored by Kyla Mandel, editor of DeSmogUK, which joins DeSmogUSA and DeSmogCanada in its aim to uncover 'the undue influence of climate science denial and the fossil fuel industry on energy and climate policy'. While *GLW* was only a small sample (n = 4), half of their articles (n = 2) were sourced from teleSUR, which is described as 'a Latin American multimedia platform oriented to lead and promote the unification of the peoples of the SOUTH. We are a space and a voice for the construction of a new communications order' (http://www.telesurtv.net/english/pages/about.html). This amplifying of voices from the global South is emblematic of the intent of these media to challenge Western hegemony.

In this study, the total absence of climate change sceptics is striking. In their stead is a long list of individuals, often contributing across Australian independent media publications and partnering with international climate change news websites. In this way, the journalists and their publications are connected both via their passionate belief in climate change and action and via circuits of climate change reporting and communication. At play here are 'both the politics of representation and politics of connectivity' (Cottle 2013: 33), characteristic of the ways in which media and transnational environmental protest and movement interact in this new media ecology. In this space, independent media and their journalists perform a role beyond reporting and representation – they are deeply and explicitly embedded in the constitution and performance of environmental issues and protest (Hutchins and Lester 2015; Lester and Cottle 2009). This is evidenced on a number of fronts: in the stated intentions of the editors, journalists and publications and their membership of various progressive social and environmental movements; their selection of commentators; and in their publication of news and comment pieces from other progressive climate change organizations. From this baseline, they set about critiquing the system and structures that hold disproportionate power in negotiating international responses at events like COP21. Alternative and independent media journalists, their commentators and the publications considered here are wired to carry out this critique and thus are providing a specific balance to climate change scepticism in the public sphere.

Sourcing practices

In relation to environmental reporting, research has consistently shown the news media's heavy reliance on authoritative sources, notably scientists and 'expert professionals, politicians and other institutional authority figures with much less prominence given to environmental pressure groups, movements and protesters' (Hansen 2015: 33). Australian research has found similar trends. In this section, I offer some analysis around the use of two key sources – Environmental Movement Organizations (EMOs) and Small Island Developing States (SIDS) – in recognition of previous research that identified their marginal status in mainstream media reporting and climate change debate, along with broader source analysis throughout. Lester *et al.* (2015: 10) investigated the presence of EMOs in reporting by the leading quality Australian daily, the *Sydney Morning Herald* (*SMH*), during successive Federal elections (1990–2013) and found an overall decline to a historical low of 4 percent in 2013. Interestingly, their research revealed a proportional decline in EMO presence in the peak years of climate change visibility in the *SMH*. Chubb's (2012: 179) analysis of Australian mainstream news reporting of COP15 and COP17 also found a near-absence of domestic NGOs or grassroots actors.

In contrast, 50 percent of the articles in *NM* and *IA* used EMOs as sources (n = 16), while at *Crikey* the proportion was even higher at just over 60 percent (n = 5). Elevating and amplifying these voices in reporting COP21 is a highly political act. These organizations are at the forefront of the challenge to hegemonic practices of 'carboniferous capitalism', as they explicitly pursue social and environmental change. There are many examples throughout the four mastheads, although *NM*'s environment writer, Thom Mitchell, is particularly exemplary on this count. Mitchell's attendance at COP21 was funded via a crowdsource campaign, adding a dimension to citizen financial support for independent media. In one of his major contributions, headlined 'Paris climate talks: the questions rich nations don't want to answer' (December 7, 2015), Mitchell led with Maina Talia, Tuvaluan activist, followed by a representative of the Climate Justice Program, Oxfam's global warming specialist, the World Wildlife Fund, the Climate Action

Network and similar others. The article canvassed proposals at COP21 to cost compensation and liability for climate change, especially in SIDS. He concludes his article:

> Given negotiations on this issue have been stilted at best so far, it seems relatively certain the future of loss and damage in the international response to climate change – and whether people like Maina Talia have a liveable future on their homelands – will be decided behind closed doors by the powerful politicians who have been dodging meaningful responsibility as emissions keep rising.
>
> (Mitchell, December 7, 2015)

News outlets in this study often prioritized and included developing nations in their COP21 reporting. This is a clear point of difference between the independent media sample here and the international experience. In contrast, Nossek and Kunelius (2012: 75–6) in their 17-nation study of COP15 and COP17 mainstream news media coverage found that reporters, even those sent to Durban for COP17, overwhelmingly quoted domestic sources. Australian mainstream media performed similarly, with domestic political voices dominating coverage (Chubb 2012: 179–80). Domestic political voices were also prevalent in this study but alongside a very strong presence of international and, in particular, political sources from SIDS. Those developing nations included were mostly from Australia's South Pacific neighbours, including Kiribati and Tuvalu. These countries do not carry the same political or economic clout in global climate change negotiations, even though they are seriously affected by climate change now. They are marginal on the world stage, so their prominence in this sample of smaller independent news sites is telling. *NM*, providing the largest collection of articles for this study, sourced 'politicians' 33 times in their coverage of COP21, and of these, just under one third (n = 10) of the sources were from developing states. Headlines from *NM* are indicative:

> December 5, 2015: Kevin Rudd Backs Pacific Island Calls for Greater Climate Action during Paris Talks
>
> December 7, 2015: Kiribati's President Is Literally Praying for Climate Success at #COP21
>
> December 12, 2015: Pacific Island Leader Slaps Down Australia's Climate Claims in Paris

Crikey also included the voices of Australia's South Pacific neighbours. Typical of the inclusion of sources from SIDS, *Crikey* journalist Karl Mathiesen (December 7, 2015) that concludes with the following:

> Unpicking the 939 square brackets [within the Paris Agreement], each of which potentially contains billions of dollars, or millions of lives, looks like child's play compared to this fundamental question of economic sovereignty.

But it must be resolved. Because if it isn't, the Paris agreement, with its recipe for 2.7 degrees of warming, will be a death warrant.

'I refuse to go home to my people without a Paris agreement that allows
me to look them in the eye and say that everything is going to be OK,' said
Tony de Brum, the Foreign Minister of the low-lying Marshall Islands on
Saturday. The stakes are intolerably high.

In their elevation of the voices of SIDS, such as the Marshall Islands' Foreign
Minister, Australian independent media empower these countries marginalized by
mainstream media and climate change policy discussions. Such media avoid the
framing of these states as 'victims', 'refugees' or 'proof of climate change' and
instead present them as authoritative voices demanding climate justice (Dreher
and Voyer 2015). The inclusion of EMOs and SIDS, alongside their various aggre-
gation and inclusion of articles from international climate change news agencies
and organizations, suggests Australia's independent media may well be performing better as a 'global citizen' through their inclusion of international sources
alongside the voices of national/domestic voices. Given the global challenge of
climate change and its local impacts on nations, regions and communities, these
media's elevation of voices marginalized by mainstream media fills a gap in the
information and ideas available for public debate and discussion.

Citizen action and critique

Citizen action and critique are strongly associated with Christians *et al.*'s (2009)
radical role and are a dominant presence in this study of Australia's independent
media. For coding purposes, 'critique' is understood more specifically as 'critical
reporting of climate change' that seeks to identify the driver or drivers of climate
change. This could involve the explicit questioning of the political, economic
and social contexts of industrial capitalism, complicit governments and industry
that profit from the continued extraction of fossil fuels. Critical reporting could
also make the link between the status quo and its deleterious consequences for
marginalized groups, communities and environments. It was agreed that the code
'citizen action' (including personal consumer decisions and protest) was defined
as either reporting on current citizen action or the promotion of citizen action in
the future. There is an obvious link between critique and citizen action. There is
little need for citizens to *do* anything if climate change reporting is framed, for
example, as a technological fix and thus the preserve of innovation and industry
which neatly folds climate change discussions back into the existing political and
economic order (McGaurr and Lester 2009). To some extent then, 'action' – in
the sense that 'something needs to be done' – is implicit to 'critique' that identifies where the problems lie. In this sample, critique appeared in more than half of
all articles. Principally, critique related to two recurrent, though often mutually
inclusive, themes in the sample: government inaction and the fossil fuel industry.
In these recurrent themes lies the heart of alternative and independent media

and their journalisms. In criticizing government inaction or inadequate action on climate change, these media hold our public offices to account. Criticism of the fossil fuel industry and/or its relations with governments has been absent in Australian mainstream media reporting (Bacon 2011, 2013; Chubb 2012), but based on this research this is a regular theme of independent media coverage of climate change.

To illustrate further, and in contrast to Bacon's findings (2011: 2013) that indicated most columnists in the Australian mainstream media were climate sceptics, the columns or commentary in the four alternative media mastheads did not promote climate scepticism. A more common orientation is found in a contribution to *Crikey*, authored by Professor Robyn Eckersley (December 12, 2015) who is a Melbourne political scientist criticizing the link between Australian government, the fossil fuel industry and climate change. Eckersley wrote:

> In short, Australia sees climate leadership as a sucker's game and climate laggardship as the smart way to operate because it believes that fossil fuels are here to stay. These are very unsafe assumptions that risk not only Australia's international reputation but also its long-term prosperity. If only Australia would look over the horizon and see what is coming: either dangerous climate change, for which we are poorly prepared, or a world that will become increasingly disinclined to buy our fossil fuel exports, which will leave us economically stranded.

Lyn Bender, in a comment piece for *IA* (December 5, 2015) was similarly scathing in her assessment of Australia's approach to the crisis faced by its Pacific neighbours:

> Like the rich man who exploits his workers and gives them a small bonus and a staff party each Xmas, the white colonial masters will throw some lifejackets to the drowning Pacific nations that have been pleading with Australia to stop mining coal. But we shall not do that. Oh no, we will continue to approve coal mines and to subsidize the fossil fuel industry.

Critique was often focused on government inaction, citing reliance on fossil fuels or the collusion between industry and governments as the cause for inertia. Repeatedly, articles made the connection between Australia's fossil fuel industry, government support and climate change. Reporting for *NM*, Mitchell (December 10, 2015) is indicative, in this case quoting the president of Kiribati to make the point in a multipronged critique of the Australian government and the fossil fuel industry:

> In October, Prime Minister Malcolm Turnbull said it 'would not make the blindest bit of difference to global emissions' if Australia stopped exporting coal because other countries would simply meet the demand.

But yesterday, the President of Kiribati, a low-lying atoll nation in the Pacific Ocean, told *NM* at the Paris climate talks that this is a 'silly argument'.

'I think what they should be doing is not doing it, and encouraging the others not to do it,' Tong said.

He was highly critical of the Australian government's response to climate change, and specifically Turnbull's dismissal of his call for a moratorium on new mines, suggesting the Coalition government 'don't feel it, they don't know it, [and] they don't care.' They care about the next election,' he said.

Citizen action was not as prominent as 'critique' throughout this sample, though this is expected given that COP21 is a U.N. conference of governments. Citizen action and protests were covered when they occurred throughout the sample period, accompanied by photos depicting protests and action. An example, again from *NM*'s Mitchell (December 12, 2015):

Activists are planning to defy a ban on political gatherings in Paris tomorrow (Saturday AEST), to highlight the need to take immediate and urgent action in a major day of civil disobedience being called D12. They'll be focusing on fossil fuels and the companies that profit from them, which have so far escaped any mention in the new international climate change regime.

And from Jonathon Neale at *GLW* (December 23, 2015) a more direct approach:

You do the math. They are lying. Emissions will rise every year. The leaders of the world have betrayed humanity. All we have on our side is seven billion people. Now we go home and mobilize.

While this research has found many distinguishing and heartening features of Australian independent media reporting on climate change, it does share some weaknesses with mainstream media. In particular, the tendency towards 'fatalism' and 'imminent disaster' in representing the climate change crisis persists in this sample. This is illustrated even across the small sample of articles cited here where climate change is variously 'a death warrant' (Mathiesen 2015), 'drowning South Pacific nations' (Bender 2015) and 'dangerous' for which we are 'poorly prepared' (Eckersley 2015). The argument is not that these are untrue assessments, but rather that these doomsday messages can lead to a sense of hopelessness and despair. In their study of Australian mainstream media outlets' coverage of climate change, McGaurr *et al.* (2013; see also Lester and McGaurr 2013) found the main messages that readers received were of 'disaster/implicit risk' or 'uncertainty'. They argue that the language of 'explicit risk' and 'opportunity' was much less prevalent, and this was true also of the other five nations[3] involved in Painter's international study (2013). Broadly, the language of 'explicit risk' implies the everyday language of insurance and decisions now to manage disasters, and 'opportunity' is understood as the possibilities that might arise in the shift to low-carbon economy and new markets (2013). On both counts and while not panacea, the language is more positive and empowering for readers than the prospect of overwhelming climate destruction and disaster.

Conclusion

The data from this study have shown an absence of climate scepticism in the four Australian alternative and independent mastheads under investigation. The intent of the journalists and the publications more broadly – in the terms of this book, the 'probable outcomes' of the journalism – enunciate a clear commitment to addressing climate change. Most of the writers possess an overt commitment to recognizing and critiquing the politics of climate change, or they are activists and leaders who are using alternative and independent media as forums for their messages. Additionally, sourcing practices elevated those marginal to the debate and most impacted – such as Small Island Developing States – to take a prominent role in coverage. At the same time, alternative and independent media provided significant space for Environmental Movement Organizations as sources, which contrasts with the approach of dominant media coverage as reported by Lester *et al.* (2015). The coverage examined here supported citizen actions and critiqued 'climate complicit and complacent' powerful governments and industry. Their focus on citizen actions was less prominent than expected – most likely because COP21 in Paris is an event for national governments. However, there was established and regular critique of governments, the fossil fuel industry and the pervading neoliberal agenda that (the journalists reported) had contributed to climate change crisis. Critique is arguably a precondition for action as it identifies problems and opportunities for change and challenge. Indeed, without 'critique' or, in this case, 'critical reporting', the status quo remains unchallenged. This research suggests that at the current historical juncture, Australian independent journalism and media are bringing diverse and unrepresented civil society voices to the public sphere and are thus performing a critical democratic role in agitating for action on climate change.

As scholars and as citizens, we know that climate change is an overwhelming and at times unfathomable challenge. A journalism is required that takes some responsibility for the communication of solutions – which perhaps represents the 'hope' with which Liz Conor from *NM* began this chapter – and connects citizen action and critique with empowerment of audiences to take part in political protest and politics. Coverage of COP21 suggests the Australian independent media explored here are some way towards delivering on this hope. In their complete silencing of climate sceptics, hope for progressive news media coverage of climate change is present. In their elevation and amplification of voices that challenge fossil fuel industries and the governments that support them, challenges to powerful institutions and structures are broadcast. The conscious inclusion of those marginal to climate change debates and discussion, but the most affected, fulfills the democratic potential of news media and journalism. The importance of these interruptions and challenges to established power – to mainstream media, the fossil fuel industry, governments and other established institutions – gains currency when climate change threatens our very survival. The performance of Australia's independent media is all the more remarkable given the preponderance of climate scepticism in mainstream media reporting and the broader historical, political, economic and cultural context of the fossil fuel industry in Australia. In the case of COP21, Australia's independent media have performed an important role in ensuring a continuing challenge to the powerful fossil fuel industry and complicit governments that hamper action on climate change.

Notes

1 Chubb's (2012) study selected Australian mainstream newspapers *The Australian, The Age* and *The Herald Sun*.
2 Data were also collected from *The Guardian Australia. The Guardian*'s reporting on climate change is the focus of Chapter Seven and is only briefly canvassed here to allow a focus on other outlets.
3 Painter's (2013) *Climate Change in the Media* investigated six nations: U.S., U.K., Norway, France, India and Australia.
4 The Hon. Malcolm Turnbull is the current Australian Prime Minister.

References

Atton, C. (2002) *Alternative Media,* London: Sage.
——— (2015) 'Introduction: Problems and positions in alternative and community media', in C. Atton (ed), *The Routledge Companion to Community and Alternative Media,* London: Routledge, pp. 1–18.
Atton, C. and Hamilton, J. (2008) *Alternative Journalism,* London: Sage.
Australian Bureau of Statistics (2012) 'Geographic distribution of population', in *Australian Year Book.* Accessed at http://www.abs.gov.au/ausstats/abs@.nsf/mf/1301.0.
Bacon, W. (2011) *A Sceptical Climate: Climate Change Policy,* Australian Centre for Independent Journalism, Sydney: UTS. Accessed at https://www.uts.edu.au/sites/default/files/sceptical-climate-part1.pdf
——— (2013) *A Sceptical Climate: Climate Science in Australian Newspapers,* Australian Centre for Independent Journalism, Sydney: UTS. Accessed at https://www.uts.edu.au/sites/default/files/Sceptical-Climate-Part-2-Climate-Science-in-Australian-Newspapers.pdf.
Bacon, W. and Nash, C. (2012) 'Playing the media game', *Journalism Studies* 13(2): 243–58.
Bender, L. (2015) 'Malcom Turnbull⁴ faking it in Paris', *Independent Australia* (5 December). Accessed at https://independentaustralia.net/politics/politics-display/malcolm-turnbull-faking-it-in-paris,8456.
Carpentier, N., Lie, R. and Servaes, J. (2003) 'Community media: Muting the democratic media discourse?', *Continuum: Journal of Media and Cultural Studies* 77(1): 52–68.
Carvalho, A. (2008) 'Media(ted) discourse and society', *Journalism Studies* 9(2): 161–77.
Carvalho, A. and Burgess, J. (2005) 'Cultural circuits of climate change in UK broadsheet newspapers,1985–2003', *Risk Analysis* 25(6): 1457–69.

Christians, C., Glasser, T., McQuail, D., Nordenstreng, K. and White, R. (2009) *Normative Theories of the Media: Journalism in Democratic Societies*, Urbana and Chicago: University of Illinois Press.

Chubb, P. (2012) '"Really, fundamentally wrong": Media coverage of the business campaign against the Australian carbon tax', in E. Eide and R. Kunelius (eds), *Media Meets Climate: The Global Challenge for Journalism*, Sweden: Nordicom, pp. 179–94.

——— (2014) *Power Failure: The Inside Story of Climate Politics Under Rudd and Gillard*, Melbourne: Black Inc.

Chubb, P. and Bacon, W. (2010) 'Fiery politics and extreme events', in E. Eide, R. Kunelius and V. Kumpu (eds), *Global Climate – Local Journalisms*, Bochum and Freiburg: Projektverlag, pp. 51–66.

Cottle, S. (2008) 'Reporting demonstrations: The changing media politics of dissent', *Media, Culture and Society* 28(2): 853–72.

——— (2013) 'Environmental conflict in a global media age: Beyond dualisms', in L. Lester and B. Hutchins (eds), *Environmental Conflict and the Media*, New York: Peter Lang, pp. 19–36.

Couldry, N. (2010) *Why Voice Matters: Culture and Politics after Neo-Liberalism*, London: Sage.

Couldry, N. and Curran, J. (2003) *Contesting Media Power: Alternative Media in a Networked World*, Lanham, MD: Rowman and Littlefield.

Downing, J.D.H. (2001) *Radical Media: Rebellious Communication and Social Movements*, Thousand Oaks, CA: Sage.

Dreher, T. and Voyer, M. (2015) 'Climate refugees or migrants? Contesting media frames on climate justice in the Pacific', *Environmental Communication* 9(1): 58–76.

Eckersley, R. (2015) 'Australia's piddling climate fund pledge leaves sour taste in Lima', *Crikey* (12 December). Accessed at https://www.crikey.com.au/2014/12/12/australias-piddling-climate-fund-pledge-leaves-sour-taste-in-lima/.

Eide, E. and Kunelius, R. (eds) (2012) *Media Meets Climate: The Global Challenge for Journalism*, Sweden: Nordicom.

Eide, E., Kunelius, R. and Kumpu, V. (2010) *Global Climate, Local Journalisms: A Transnational Study of How Media Make Sense of Climate Change Summits*, Bochum and Freiburg: Projektverlag.

Forde, S. (2011) *Challenging the News: The Journalism of Alternative and Community Media*, London: Palgrave Macmillan.

——— (2015) 'Politics, participation and the people: Alternative journalism around the world', in C. Atton (ed), *The Routledge Companion to Community and Alternative Media*, London: Routledge, pp. 291–300.

Forde, S., Foxwell, K. and Meadows, M. (2009) *Developing Dialogues: Indigenous and Ethnic Community Broadcasting in Australia*, United Kingdom: Intellect Publishing, and Chicago: University of Chicago Press.

Foxwell-Norton, K. (2015) 'Community and alternative media: Prospects for 21st century environmental issues', in C. Atton (ed), *The Routledge Companion to Community and Alternative Media*, London: Routledge, pp. 389–400.

——— (2016) 'The global alternative and community media sector: Prospects in an era of climate crisis', Editorial Contribution, *Journal of Alternative and Community Media*. Accessed at https://joacm.org/index.php/joacm/article/view/850.

Fuchs, C. and Sandoval, M. (2015) 'The political economy of capitalist and alternative social media', in C. Atton (ed), *The Routledge Companion to Community and Alternative Media*, London: Routledge, pp. 165–76.

Garnaut, R. (2008) 'Garnaut climate change review: Interim report to the Commonwealth, state and territory governments of Australia', in *Technical Report*, Melbourne: Garnaut Review Secretariat. Accessed at http://eprints3.cipd.esrc.unimelb.edu.au/392/.

Garnett, A. (2015) 'Australia's "five pillar economy": Mining', *The Conversation* (May 1). Accessed at https://theconversation.com/australias-five-pillar-economy-mining-40701.

Graham, C. (2016) 'Two years today: Keeping the faith in a torrid independent media love affair', *New Matilda* (May 19). Accessed at https://newmatilda.com/2016/05/19/two-years-today-keeping-the-faith-in-a-torrid-independent-media-love-affair/.

Gunster, S. (2011) 'Covering Copenhagen: Climate change in BC media', *Canadian Journal of Communication* 36(3): 477–502.

——— (2012) 'Visions of climate politics in alternative media', in A. Carvalho and T.R. Peterson (eds), *Climate Change Politics: Communication and Public Engagement*, Cambria Press: New York, pp. 247–77.

Gurney, M. (2013) 'Whither the "moral imperative"? The focus and framing of political rhetoric in the climate change debate in Australia', in L. Lester and B. Hutchins (eds), *Environmental Conflict and the Media*, New York: Peter Lang, pp. 187–200.

Hackett, R.A. (2000) 'Taking back the media: Notes on the potential for a communicative democracy movement', *Studies in Political Economy* 63: 61–86.

Hansen, A. (2015) 'Communication, media and the social construction of the environment', in A. Hansen and R. Cox (eds), *The Routledge Handbook of Environment and Communication*, London: Routledge, pp. 26–38.

Hopke, J.E. (2012) 'Water gives life: Framing an environmental justice movement in the mainstream and alternative Salvadoran press', *Environmental Communication: A Journal of Nature and Culture* 6(3): 365–82.

Howley, K. (2005) *Community Media: People, Places and Communication Technologies*, New York: Cambridge University Press.

Hsieh, H.-F. and Shannon, S.E. (2005) 'Three approaches to qualitative content analysis', *Qualitative Health Research* 15(9): 1277–88.

Hulme, M. (2009) *Why We Disagree about Climate Change: Understanding Controversy, Inaction and Opportunity*, Cambridge, UK: Cambridge University Press.

——— (2015) '(Still) disagreeing about climate change: Which way forward?', *Zygon* 50(4): 893–905.

Hutchins, B. and Lester, L. (2015) 'Theorizing the enactment of mediatized environmental conflict', *International Communication Gazette* 77(4): 337–58.

Jackson, E. (2015) 'Day 1, Paris climate talks: Helping dirty power get clean', *Crikey* (1 December). Accessed at https://www.crikey.com.au/2015/12/01/day-1-paris-climate-talks-helping-dirty-power-get-clean/.

——— (2015) 'Memo to Paris climate delegates: Compromise now or world is fukt', *Crikey* (8 December). Accessed at https://www.crikey.com.au/2015/12/08/memo-to-paris-climate-delegates-compromise-now-or-world-is-fukt/.

——— (2015) 'Sleepless nights and saving the world in Paris', *Crikey* (11 December). Accessed at https://www.crikey.com.au/2015/12/11/sleepless-nights-and-saving-the-world-in-paris/.

Kenix, L.J. (2008) 'Framing science: Climate change in the mainstream and alternative news of New Zealand', *Political Science*, Special issue: *The Politics of Climate Change: Issues for New Zealand and Small States of the Pacific* 60(1): 117–32. Accessed at http://dx.doi.org/10.1177/003231870806000110.

Laclau, E. and Mouffe, C. (1985) *Hegemony and Socialist Strategy: Towards a Radical Democratic Politics*, London: Verso.

Lester, L. (2010) *Media and Environment*, Cambridge, UK: Polity Press.

——— (2015) 'Containment and reach: The changing ecology of environmental communication: News and new media roles', in A. Hansen and R. Cox (eds), *Routledge Handbook of Environmental Communication*, Abingdon, UK: Routledge, pp. 232–41.

Lester, L. and Cottle, S. (2009) 'Visualising climate change: TV news and ecological citizenship', *International Journal of Communication* 3: 920–36.

Lester, L. and McGaurr, L. (2013) 'Australia', in J. Painter (ed), *Climate Change in the Media: Reporting Risk and Uncertainty*, Oxford, UK: RISJ, pp. 79–88.

Lester, L., McGaurr, L. and Tranter, B. (2015) 'The election that forgot the environment? Issues, EMOs, and the press in Australia', *The International Journal of Press/Politics* 20(1): 3–25.

Leviston, Z., Greenhill, M. and Walker, I. (2015) *Australians Attitudes to Climate Change and Adaptation: 2010–2014*, Australia: CSIRO.

Mandel, K. (2015) 'Malcom Turnbull implementing Tony Abbott's policies #COP21 says Shorten', *Independent Australia* (1 December). Accessed at https://independent australia.net/environment/environment-display/turnbull-implementing-tony-abbotts-policies-at-cop21-says-shorten,8440.

Mathiesen, K. (2015) 'Now comes the diplomatic bastardry: Paris climate delegates find 939 things to fight about', *Crikey* (7 December). Accessed at https://www.crikey.com.au/2015/12/07/now-comes-the-diplomatic-bastardry-paris-climate-delegates-find-939-things-to-fight-about/.

Mayring, P. (2000) 'Qualitative content analysis [28 paragraphs]', *Forum: Qualitative Sozialforschung / Forum: Qualitative Social Research* 1(2): Art. 20. Accessed at http://nbnresolving.de/urn:nbn:de:0114-fqs0002204.

McGaurr, L. and Lester, L. (2009) 'Complementary problems, competing risks: Climate change, nuclear energy and the Australian', in T. Boyce and J. Lewis (eds), *Climate Change and the Media*, New York: Peter Lang, pp. 174–85.

McGaurr, L., Lester, L. and Painter, J. (2013) 'Risk, uncertainty and opportunity in climate change coverage: Australia compared', *Australian Journalism Review* 35(2): 21–33.

McKnight, D. (2010) 'A change in the climate? The journalism of opinion at News Corporation', *Journalism* 11(6): 693–706.

Mitchell, T. (2015) 'Paris climate talks: The question rich nations don't want to answer', *New Matilda* (7 December). Accessed at https://newmatilda.com/2015/12/12/paris-climate-talks-good-deal-still-possible-but-not-assured/.

——— (2015) 'Activists move to shut down east coast coal exports in coordinated action', *New Matilda* (10 December). Accessed at https://newmatilda.com/2015/12/10/activists-move-to-shut-down-east-coast-coal-exports-in-coordinated-action/.

——— (2015) 'Paris climate talks: Good deal still possible, but not assured', *New Matilda* (12 December). Accessed at https://newmatilda.com/2015/12/12/paris-climate-talks-good-deal-still-possible-but-not-assured/.

Moser, S.C. (2015) 'Whither the heart(-to-heart)? Prospects for a humanistic turn in environmental communication as the world changes darkly', in A. Hansen and R. Cox (eds), *The Handbook of Environment and Communication*, London: Routledge, pp. 402–13.

Mouffe, C. (2013) *Agonistics: Thinking the World Politically*, London: Verso.

Neale, J. (2015) 'Why the COP21 agreement will raise carbon emissions', *Green Left Weekly* (23 December). Accessed at https://www.greenleft.org.au/content/why-cop21-agreement-will-raise-carbon-emissions.

Nossek, H. and Kunelius, R. (2012) 'News flows, global journalism and climate summits', in E. Eide and R. Kunelius (eds), *Media Meets Climate: The Global Challenge for Journalism*, Sweden: Nordicom, pp. 67–86.

Painter, J. (2013) *Climate Change in the Media: Reporting Risk and Uncertainty*, Oxford, UK: RISJ.

Parkinson, G. (2015) 'Paris, COP21: Poor countries want renewables not coal', *Independent Australia* (3 December). Accessed at https://independentaustralia.net/politics/politics-display/paris-cop21-poor-countries-want-100-per-cent-renewables-not-coal,8448.

Reese, S. (1994) 'The structure of news sources on television: A network analysis of CBC News, Nightline, MacNeil/Lehrer, and This Week with David Brinkley', *Journal of Communication* 44(2): 84–107.

Rodriguez, C. (2001) *Fissures in the Mediascape: An International Study of Citizens Media*, Cresskill, NJ: Hampton Press.

——— (2011) *Citizens' Media Against Armed Conflict: Disrupting Violence in Colombia*, Minneapolis: The University of Minnesota Press.

Schreier, M. (2014) 'Qualitative content analysis', in U. Flick (ed), *The SAGE Handbook of Qualitative Data Analysis*, London: Sage Publications, pp. 170–83.

Smeltzer, S. (2008) 'Biotechnology, the environment, and alternative media in Malaysia', *Canadian Journal of Communication* 33(1): 5–20.

Tranter, B. and Booth, K. (2015) 'Scepticism in a changing climate: A cross-national study', *Global Environmental Change* 33: 154–64.

van Dijk, T. (2000) 'New(s) racism: A discourse analytical approach', in S. Cottle (ed), *Ethnic Minorities and the Media: Changing Cultural Boundaries*, Buckingham, UK: Open University Press, pp. 33–49.

Walsh, K. (2013) *The Stalking of Julia Gillard: How the Media and Team Rudd Brought Down the Prime Minister*, Sydney: Allen & Unwin.

Chapter 7

Alternative approaches to environment coverage in the digital era

The Guardian's 'Keep it in the Ground' campaign

Susan Forde

Introduction: The not-so-new media

In a new offering which outlines media theories and practices in the digital age, Bainbridge *et al.* (2015) note that while Internet-based media have in the past been referred to simply as 'new media', their proliferation and complete integration into the media landscape means they are now more specifically referred to as 'digital media' and 'social media' (p. 3). Indeed, these media forms based entirely on electronic 'outputs' are sometimes recognized as 'The Fifth Estate', a new addition to the institutions of the modern world, and to flag the advance that they offer on the media as the 'Fourth Estate' (Bainbridge *et al.* 2015). Flew (2008) notes that the so-called 'new media' comprise 'the three Cs – computing and information technology (IT), communications networks and digitized media and information content – arising out of another process beginning with a "C", that of convergence' (p. 3). News consumers can now participate in generating news stories from data about, for example, climate change impacts or the measurement of countries' carbon footprints – a politically loaded issue in the context of emerging global emission ceilings and demands for climate justice. Similar technology enables the transformation of such data into visual images that could make climate change far more understandable to broad publics and link global patterns to local situations. One example in the environmental field is the 2010 *New York Times* project on worsening pollution in American waters and regulators' responses. The key story – that 62 million Americans had been exposed to toxic tap water since 2004 – was supplemented with interactive graphics and quantitative data from searchable databases that enabled readers to check out the situation in their own city or state. Attracting millions of clicks, the project led to tougher environmental regulation and new spending on clean-ups (Gynnild 2014: 719–20).

This anecdote leads us to question whether the action occurred as a result of the 'new' way the information was presented – in accessible and interactive digital graphics – or simply because the story was impossible to ignore and created considerable public concern which led to changed legislation. Lister *et al.* (2009)

confirm that 'speculation, prediction, theorization and argument about the nature and potential of these new media began to proceed at a bewildering and breathless pace', with past and present research challenging existing assumptions about media, culture, technology and nature (p. 2). The emergence of new media in the mid-1980s was greeted with hype, as is the case with every new technological development (see also Flew 2008), but the field is now in a position to apply some 'hard-headed reflection born of experience and [with] enough time to recover some critical poise' (Lister *et al.* 2009: 2). They suggest, however, that in contrast with techno-utopian hype, strictly critical accounts of the impact of digital media may sell the new media short, minimizing their impact by identifying new media's position as part of the 'capitalist scam':

> Such critical accounts of new media frequently stress the continuity in economic interests, political imperatives and cultural values that drive and shape the 'new' as much as the 'old' media. . . . They argue that new media can largely be revealed as the latest twist in capitalism's ruthless ingenuity for ripping us off with seductive commodities and the false promise of a better life.
>
> (p. 3)

This book does not take this somewhat simplistic approach summarized so bluntly by Lister *et al.*, although we support a tempering of the hype surrounding the emancipation that digital media have purportedly delivered to democracies, publics and radical social movements. Authors such as Cammaerts *et al.* (2013), Radsch (2011), Sreberny (2011) and Wolfsfeld *et al.* (2013) have variously focused on the impact social media have had on revolutionary protest movements such as the Arab Spring, the Occupy protests and the Indignado in Spain. Earlier work focused on the development of the Indymedia concept emerging from the World Trade Organization protests in Seattle in 1999 and representing one of the first open-source alternative media experiments (see Atton 2007; Hyde 2002; Platon and Deuze 2003). Hyde in particular writes in *First Monday*, the peer-reviewed journal about the Internet, promoting the democratizing potential of Indymedia and its growth from one Independent Media Center in 1999 to more than 60 around the world by 2002. An Indymedia journalist whom Hyde interviewed told him: 'We had a saying in Seattle. . . . We're trying to break through the information blockade.' This 'information blockade' referred to the attempts by corporate-owned media to misrepresent or to simply fail to report the 'varied viewpoints of those who protest globalization and the WTO' (Hyde 2002). Radical media theorist John Downing recognized the empowering potential of Indymedia in 2003, although some reflections identified the vulnerabilities of Indymedia's open-publishing system which saw many right-wing groups, homophobes and racists publish content on the Indymedia sites (Atton 2003).

This chapter, then, provides an overview of the impact of digital and social media on journalism and social movements. It orients the focus specifically to environmental crisis and climate change. The aim of the chapter is to consider in-depth a particular journalistic model – demonstrated in *The Guardian*'s 'Keep it in the Ground' anti–fossil fuels campaign – and to evaluate the motivations and editorial processes which contributed to this particular journalistic experiment on environmental reporting. Before the case study is introduced, I provide some background to both alternative and mainstream journalism's interactions with digital and social media.

Digital media, social media and journalism

Research on the impact of digital and social media on journalism swings widely between a fairly common view that new media have revolutionized communication and cautious assessments that see new media as simply a new tool for journalists, editors, activists and broader social movements to use to best effect. Juris (2005) argues that independent media activists 'have made particularly effective use of new technologies through alternative and tactical forms of digital media production' (p. 200). He finds digital technologies have allowed anti-globalization activists to sidestep the mass media to circulate messages:

> While in the past activists had to rely on experts and the mass media to circulate their messages, largely due to high transaction costs and time constraints, they can now use new digital technologies to take on much of this work themselves, assuming greater control over the media production process, while enhancing the speed of information flow.
>
> (p. 201)

He finds the 'temporary media labs' established by activist groups around technology and digital media opportunities had facilitated the exchange of information, ideas and resources, 'as well as experimentation with new digital technologies through which media activists inscribe their emerging political ideals within new forms of networked space'. Hutchins and Lester's (2015) discussion of environmental protest and associated media coverage in the Upper Florentine Valley in the southwest of Tasmania describes the effective combination of on-site protest with the online distribution of video footage through YouTube. Dramatic footage – of two protesters chained inside a wrecked car body that is then attacked by loggers unaware they were being filmed – received widespread coverage from major media. Hutchins and Lester particularly refer to the 'switching point' in this mediatized environmental conflict, the moment when the protest group – Still Wild, Still Threatened – uploaded the footage to both MySpace and YouTube:

> This mechanism shifted the dominant public account of the incident in the favor of activists, transforming the episode from an 'isolated incident' in a remote forest into a 'savage attack' on peaceful protestors. . . . The shocking

nature of the footage and viral distribution of the video obligated both government and industry representatives to answer difficult questions posed in the news media.

(p. 346)

Beyond easily accessible online video sites, another growing digital content form that has affected journalism is blogging. Marcus Leaning (2011) notes that technological determinism suggests blogs will transform the way journalism is practised and the way people engage with media (p. 92). While in traditional media systems citizens are broadcast *to,* blogs and other forms of new technology 'empower the citizen in the face of corporate and government control':

> Blogs are of course not the first media form credited with potential to radically challenge the status quo. Indeed, it seems almost a 'rite of passage' for any new media form or technology to be considered dangerous, radical or subversive.
>
> (Leaning 2011: 88, citing Carey 1989)

Leaning notes, however, that most bloggers do not consider themselves journalists; while some blogs offer a form of citizen journalism, many are simply republishing content from other (usually mainstream) media outlets. Importantly, audiences for blogs are notoriously partisan and tend to follow bloggers that they intensely agree with; essentially, bloggers speak to a particular community of interest and do not necessarily contribute substantially to broader public debate (2011: 97). And while Jürgen Habermas (2006) sees real benefits for publics in repressive regimes from blogs, chat rooms and the Internet more generally, in liberal societies he only sees

> the rise of millions of fragmented chat rooms across the world [which] tend instead to lead to the fragmentation of large but politically focused mass audiences into a huge number of isolated issue publics.
>
> (p. 423)

This certainly confirms Leaning's notion – drawn from broad literature surveys – that many blogging and Internet-based news communications are targeted at a niche, already supportive audience. There is value in such targeting for a social movement like environmentalism – it is a way to confirm the breadth of support for the ideas at hand and to build community, which is an identified strength of community radio (see e.g. Anderson 2012; Meadows *et al.* 2007 discusses this in the context of building 'community' among prisoners). Foxwell-Norton (2015) identifies the central place that local communities occupy in forming issue publics around environmental issues: 'communities do not just live in their local environment; they also "think" their environments, bringing a host of cultural frameworks to their place' (p. 393).

Bradshaw and Rohumaa's (2011) guide to online reporting and data journalism for budding journalists offers excitement about the potential for 'new stories' that data journalism brings, expanding the possible ways of delivering particular stories because of the new content and depth of analysis that it makes possible (p. 71). A key feature of data journalism relevant to this work is its enhanced opportunity to engage 'users' (in our terms, publics) with the news in ways that legacy media cannot. Audiences are invited to examine data, to find new ways to analyze it, to participate in the newsmaking process – 'it is a newsroom without walls' (Bradshaw and Rohumaa 2011). Of course, this notion has a great deal in common with the well-established practices of community media, and research has long acknowledged the collapsing of the audience–producer boundary in many community and grassroots media outlets (Atton 2002; Meadows *et al.* 2007; Rennie 2006; Rodriguez 2001). There is nothing new in this, although it is a relatively new idea – engaging properly with audiences and inviting their participation – for mainstream media outlets. Further, the more widespread use of data journalism and the increasing capacity of organizations to collect and analyze 'big data' gives enhanced opportunity for social movements and their media to delve into government and corporate records for stories. Community and alternative media journalists will have particular motivation to do this if they see potential exposés being ignored by the mainstream.

Bradshaw (in Bradshaw and Rohumaa 2011) suggests data journalism may be a 'new genre', even though it has much in common with the 1990s concept of 'computer-assisted reporting' (CAR). Data journalism occurs in a more efficient form which gives access to larger sets of data. Still, CAR was a new concept at the time, one which encouraged journalists to deliver original analysis to company spreadsheets and annual reports through the use of new computer software (such as Excel and analytical software such as SPSS, etc). The United States' Poynter Institute, a leading think tank and research institute for investigations about the news media and journalism, encouraged its members in 2005 to engage in computer-assisted reporting because 'a database, a spreadsheet, helps us get some stories that just can't be acquired any other way' and they help journalists 'add depth and detail that fascinate readers' (Stith 2005). Stith discusses the ways that CAR can enhance the Fourth Estate role of the media by providing an enhanced ability to act as a watchdog on power, for example. Journalistic access to big data has the potential to deliver new and empowering information to readers in the same way that social media and digital media have the potential to activate large numbers of people on major social issues – in our case, climate change. Whether this potential is realized is a more complex matter and is affected by political-economic factors, limited journalistic rituals, commercial considerations and the ability of individual journalists and their outlets to be resourced sufficiently to conduct the 'big data' research in the first place. Data journalism's democratizing impact may well be constrained by its dependence on well-resourced organizations, trained professionals and the accessibility and accuracy of large-scale, institutionally generated databanks, which could 'serve certain political or ideological interests' (Lesage and Hackett 2013: 47).

Bruns (2015) draws connections between the development of the practices of citizen journalism and the technological frameworks of the Web 2.0 publishing technologies of news blogs, open-source Internet publishing and social media, all of which are underpinned by online platforms. He is cautious, however, that drawing this connection 'is by no means to fall into the trap of technological determinism: in a variety of forms, and without using the term itself, citizen journalism has been practiced . . . for decades, even centuries' (p. 379) and that 'the emerging Web 2.0 simply became the latest and particularly powerful set of tools with which to engage in an alternative form of newswriting and commentary – tools so versatile that the term 'citizen journalism' itself was born' (pp. 379–80).

Tellingly, James Carey wrote in 1969 of the 'communication revolution', the development of the telegraph, the growth of the mass media, conduits of information which 'cut across structural divisions in society. . . . Modern communications media allowed individuals to be linked, for the first time, directly to a national community without the mediating influence of regional and other local affiliations' (Carey 1969: 129). He sees an important characteristic of this communication revolution in the 'development of specialized media of communication located in ethnic, occupational, class, regional, religious and other "special interest" segments of society' and that these minority media 'are in many ways more crucial forms of communication because they are building blocks upon which the social structure is built up and they serve as intermediate mechanisms linking local and partial milieus to the wider community' (p. 130). These media represent a 'centrifugal force in social organization' by, among other things, 'transforming groups into audiences' (p. 131). Why introduce Carey's 1969 work here? In an effort to posit that this *new* communications revolution – which we could easily get the feeling is the first, the only, the most transformative – has happened before and affected societies, politics and culture. In accepting this, we are able to see digital technologies as a new phase, a more powerful way to communicate, but one which will not necessarily transform existing structures in any radical way in the absence of real, structural change to our 'system'.

Poell and Borra (2012) have used the controversial G20 Toronto protests in 2009 to assess the impact of digital media and social media portals on the practice of alternative journalism. Their aim is to determine if the impact and reach of alternative media practitioners has extended and indeed become more effective as a result of digital technologies. They particularly examined Twitter, Flickr and YouTube – a combined sample of both digital and social media encompassing the Web 2.0 advances. Their analysis of the social media content of alternative media during the G20 protests indicated that alternative media, when using such social media platforms, often delivered similar content to the mainstream media. Essentially, social media forms forced alternative media practitioners to deliver event-driven content that focused on police conflict rather than a more considered treatment of the issues which usually characterized alternative media coverage (p. 705). They gathered their data by monitoring the alternative media's hashtag #g20report, and this confirmed that the content was events- and conflict-focused rather than revealing the reasons behind the protests which might encourage reader action (2012: 705ff). Similarly Wolfsfeld *et al.*, in examining the Arab Spring coverage through social media and assessing the impact of social media on activism and political effectiveness, found 'a consistently negative correlation between the extent of social media penetration and the amount of protests' (2013: 132). Their overall findings were that much research seems to 'overemphasize the centrality of social media in protest. As always the "real" question is not whether this or that type of media plays a major role but how that role varies over time and circumstance' (2013: 132). Lance Bennett (2003) reminds us that it is the context of the technology, not the technology itself, that should be considered if we are to assess the empowerment potential of the Internet and, by extension, digital and social media (p. 19). He warns that the rise in global activism which is often presented as a result of the capabilities of new technologies and portals (such as Facebook, Twitter, Weibo in China, Tumblr and YouTube) cannot be wholly attributed to the reduced costs of the Internet and its potential to coordinate activism and collective protest. Rather it is the 'social and political dynamics of protest' that have changed 'due to the ways in which economic globalization has refigured politics, social institutions and identity formation within societies' (p. 25). Gerbaudo (2012) suggests that 'social media can be seen as the contemporary equivalent of what the newspaper, the poster, the leaflet or direct mail were for the labour movement' (p. 4). They are now primarily a way to notify activists of events and to encourage their involvement. There is little doubt that the potential networking and campaign opportunities that social media provide will form part of the work – perhaps a core part – of alternative and community media journalists and producers from this point on. However, the evidence suggests that particular forms of social media might have a real – and negative – impact on alternative journalism in that they force content into a conflict-/events-driven agenda; and while intuitively social media's impact on protest action is positive due to their broad reach and efficiency, the actual evidence is mixed.

Case study in climate change journalism: *The Guardian*

I will now consider the content of an independent journalism website – *The Guardian Australia* – to look for signs of the type of journalisms we might consider useful to climate change reporting. *The Guardian* (and note, I have relied on the 'Australian edition' for data collection, although the content was emanating from *The Guardian*'s U.K. headquarters) was chosen primarily because it appears to be driven by a few converging and complementary factors: it is seeking to distinguish itself from other daily media in Australia by presenting a progressive view of current issues; it is identifying the weaknesses in current mainstream media coverage and practices and trying to show 'a new way' for journalism; and its ownership structure is a trust which regularly reaffirms its commitment to producing a strong journalistic product, rather than maximizing profit and readership.

Importantly for this particular chapter, *The Guardian* (Australia) is an entirely online newspaper, and there is little doubt that in the pre-digital era it would not be publishing a daily news product. It is the efficiencies of the Internet – both in publishing content and disseminating it to a wide audience – that has facilitated the growth of a range of online alternative and independent news sites, and *The Guardian* is a key example.

I have chosen a case study approach for this work to be able to highlight very specific practices, to examine them in detail and to track through their impact. Following Flyvberg (2004), this chapter provides 'a detailed examination of a single example' (p. 420) and is designed to interrogate the editorial and journalistic tone and practice of *The Guardian*'s 'Keep it in the Ground' (KIITG) campaign in detail. It is important in case study research to ensure the data gathering goes beyond 'a method of producing anecdotes' and arrives at considered conclusions based on in-depth examples (p. 422). The data drawn from *The Guardian* for this chapter is indeed an 'information-oriented selection' (p. 426) designed to illustrate some of the key themes discussed in this book. Specifically, I examined the initial articles from both *The Guardian*'s editor, Alan Rusbridger, and *The Guardian* journalists at the launch of the campaign to understand the rationale behind it and to establish the publication's motivations and objectives. I also examined a selection of copy during the campaign which overtly speaks of the journalism undertaken, supplemented by analysis of the 12 podcast episodes of the KIITG campaign which feature the 'behind-the-scenes' activity of journalists and editors in delivering this 'new way' of reporting climate change. Some lengthy extracts from the publication have been quoted to illustrate clearly the rationale for the practice and to assess fully the journalistic shifts that might be present.

To start with, a basic walk through home pages of a range of Australian news sites indicates *The Guardian* is already well ahead in terms of coverage of the environment – out of five major news sites, it is the only one to include 'Environment' as one of its tabs on the news home page. That is, it was the only news site that indicated to readers the publication had a particular reporter or 'round' dedicated to environment coverage. All other major Australian sites – the leading news.com.au, which is an aggregate of all News Corp Australia (Murdoch) newspapers; the national newspaper *The Australian* (also owned by Murdoch); the Fairfax-owned leading quality daily *The Sydney Morning Herald*; and the highest-selling Australian newspaper, another Murdoch publication, *The Daily Telegraph* – did *not* indicate a dedicated section to environment coverage in 2016. Their publications included, in all cases, variations on news categories of National, Politics, Business, Sport, Travel, Entertainment, Technology, Lifestyle and so on, but no Environment. On finding the Environment tab on *The Guardian*'s home page, readers are directed to a series of news, feature and comment pieces which cover climate change, coral bleaching in the Great Barrier Reef and the impact of the coal seam gas industry, along with a series of nature-based photo essays.

Identifying the environment as a major news round for the publication immediately prioritizes this issue for readers, and its exclusion from other outlets communicates a similar, but opposite, message.

The Guardian's 'Keep it in the Ground' campaign

In March 2015, *The Guardian* online newspaper told its readers:

> Climate change is the biggest story journalism has never successfully told. *The Guardian*'s editor-in-chief, Alan Rusbridger, has decided to change that. This podcast series follows Rusbridger and his team as they set out to find a new narrative on the greatest threat to humanity (The Guardian, Thursday March 12, 2015).

This story in *The Guardian*, headlined 'Find a new way to tell the story', introduces readers to the 'Keep it in the Ground' campaign, the newspaper's attempt to encourage mining companies to leave fossil fuels in their place, in the ground. The main focus of this first piece is to enable readers to take the journey with reporters from *The Guardian* to find a new way to tell the climate change story – to listen to podcasts from the newsroom and to see how journalists manage the process, the mistakes they make, the successes they have (Krotoski 2015). KIITG was launched as a proactive environmental campaign by *The Guardian* newspaper in March 2015 to encourage a range of action from major corporate investors, governments and citizens to take part in a movement to reduce the use of coal, oil and gas reserves. KIITG is an international news-media driven initiative, with former *Guardian* Editor-in-Chief Alan Rusbridger launching the campaign in London. The idea to 'keep it in the ground' comes from environmentalist Bill McKibben, founder of the 350.org social movement and leading international environmentalist, journalist and author of 25 books, including *The End of Nature* (1989), the work that flagged global warming and the necessary change to human behaviour in order to salvage a sustainable future.

Rusbridger (2015) identifies the unique but well-precedented genre of journalism that takes a stand:

> The usual rule of newspaper campaigns is that you don't start one unless you know you're going to win it. This one will almost certainly be won in time: the physics is unarguable. But we are launching our campaign today in the firm belief that it will force the issue now into the boardrooms and inboxes of people who have millions of dollars at their disposal.

The campaign focuses on divestment, and it began with a plea to major investment trusts the Bill and Melinda Gates Foundation and the Wellcome Trust to divest their investments in the top 200 companies involved in the fossil fuels industry. *The Guardian*'s focus is squarely on the role journalism can and should play in not just providing information and creating awareness but in animating the public to take action. Crucially for our book's argument, this content provides a model for an alternative form of journalism which, in the face of major crises, jumps off the fence and takes responsibility for playing a key public role in facilitating debate and encouraging and enabling citizen action. The KIITG campaign was the culmination of professional discussions among the board, editorial chiefs and journalistic staff of *The Guardian*. Rusbridger tells readers:

> When, as *Guardian* colleagues, we first started discussing this climate change series, there were advocates for focusing the main attention on governments. States own much of the fossil fuels that can never be allowed to be dug up. Only states, it was argued, can forge the treaties that count. In the end the politicians will have to save us through regulation – either by limiting the amount of stuff that is extracted, or else by taxing, pricing and limiting the carbon that's burned.
>
> If journalism has so far failed to animate the public to exert sufficient pressure on politics through reporting and analysis, it seemed doubtful whether many people would be motivated by the idea of campaigning for a paragraph to be inserted into the negotiating text at the UN climate talks in Paris this December. So we turned to an area where campaigners have recently begun to have marked successes: divestment.

The online newspaper campaign contains news articles, longer-form feature articles, opinion pieces from climate change scientists and advocates such as McKibben, alongside compelling still photography and video content which visualizes the climate crisis. Some of the video content (called 'video explainers' and put together by *Guardian* journalists) links the techniques used in the KIITG campaign back to the calls from South African anti-apartheid figures Nelson Mandela and Archbishop Desmond Tutu who called on the international monetary community to divest ownership in major South African corporations in order to demonstrate their opposition to the apartheid regime. Sources in *The Guardian*'s video content identify that 'a generation ago', apartheid was the biggest moral issue facing the international community; at that time, South African leaders suggested that 'it was time for the great institutions of the West to cut their economic ties with the companies that propped up the apartheid regime.' (Poulton, L et al., 2015) The audio from the interviews runs over the top of images from the worst period of the apartheid era, with black South African protesters being dispersed with high-powered water hoses and the slums of Johannesburg, evoking the site of the Sharpeville massacre. In drawing connections with the tactics of the anti-apartheid campaigners, *The Guardian* is attempting to place its KIITG campaign in a continuum of moral and ethical fights against recognized 'wrongs' in the world – in this case, the continued extraction of fossil fuels for economic gain with little regard for the future of humanity and the planet. It is not a human rights campaign in the same way the anti-apartheid movement was, but it is identified as the major moral issue facing the international community at this point in human history (*The Guardian*, March 15, 2015; video contained in Rusbridger's editorial). In short, 'unless there is political pressure to change, nothing will change.' (Poulton, L et al., 2015)

The Guardian then goes further, urging personnel currently involved in the fossil fuel industry to contact Rusbridger directly if they wish to confidentially pass on information that might enable the newspaper to further 'fuel' their campaign. Here, the newspaper is flagging a need to move well outside the usual sources to include the environmental movement and other progressive social movements, as well as whistleblowers or dissidents in industry and government. Rusbridger finishes his editorial to launch KIITG as follows:

> One final thing. This campaign is going to be backed up by much reporting and analysis. We would be very pleased to hear from anyone working in the fossil fuel industries at a senior level, either currently or recently. We are interested, for instance, to learn about internal discussions and papers about the state of knowledge and debate about the environmental harm caused by the extractive industries. You can email me confidentially at alan.rusbridger@ theguardian.com; see my PGP key on @arusbridger on Twitter; or use *The Guardian*'s encrypted securedrop platform, which enables anyone to send us documents without being traced.

Phase II of KIITG launched in October 2015. *The Guardian* explained their rationale and future direction as follows (Randerson 2015):

> A year ago, more than 300,000 people took to the streets in New York to demand action from their leaders on climate change.
>
> Nearly the same number took part in similar events in 161 countries across the globe. For 24 hours, the sun did not set on the largest climate protest in history.
>
> These grassroots activists are part of a powerful global movement for change that has continued to grow as crucial UN climate talks in Paris in December have drawn nearer, bolstered by interventions from other important global voices – Pope Francis, Graça Machel, Desmond Tutu and Mary Robinson, to name a few. The Pope last week repeated his message of climate justice and change to world leaders at the UN.
>
> Crucially, that change is now beginning to take hold, with clean energy on the march and the low-carbon economy becoming a reality on the ground, rather than just a PowerPoint aspiration.
>
> It is against this backdrop that *The Guardian* is launching the next stage of its climate change campaign as our team of environment correspondents around the world champion a rare commodity in the climate change debate – hope.

The headline to the launch of Phase II of the KIITG campaign pledges 'A story of hope: *The Guardian* launches Phase II of its climate change campaign' with the major caption to the page declaring: *The next phase of our campaign will champion solar power and its potential to transform the global energy supply.* The use of the terms 'hope' and 'champion' overtly flags the advocacy nature of the campaign and the implied agency that it delivers to readers, in that it represents a source of 'hope' and the potential to cause change. The sub-heading calls to readers:

> With crucial climate talks on the horizon, *Keep it in the Ground* turns its focus to hope for the future – the power to change and the solar revolution. Join us and help make that change happen. (Randerson 2015)

It further encourages: 'Solar energy is taking off around the world. Join *The Guardian*'s campaign and help spread a message of hope that the world can stop climate change.'

The Guardian's commitment is backed by its own board's decision in April 2015 to divest more than €800 million from coal, oil and gas interests, with chairman of *The Guardian*'s board, Neil Berkett, declaring that the Guardian Media Group (GMG) became the largest fund yet to pull out of coal, gas and oil company investment in a decision that Berkett called a 'hard-nosed business decision' justified on ethical and financial grounds (Carrington, 2015). The opportunities for readers to participate in the KIITG campaign and to take action and affect change are consistently reinforced by columnists, journalists and editors alike in the KIITG pages:

> So whether you are already a supporter of Keep it in the Ground or whether you are seeing the *Guardian*'s campaign for the first time, please sign up to find out more. By doing so, you will receive regular updates on our coverage and the progress of the campaign, as well as an opportunity to participate and influence the direction we take.
>
> This is the most exciting and hopeful time for anyone interested in solving the biggest problem that humanity faces. As Pope Francis put it in his encyclical on the environment in June: 'All is not lost. Human beings, while capable of the worst, are also capable of rising above themselves, choosing again what is good, and making a new start . . . to embark on new paths to authentic freedom.'
>
> That new start is already rising from the dirty energy system we inherited from the 19th and 20th centuries but for now it is just that – a start. It is only with unrelenting pressure from below that world leaders will continue with enough purpose on the right path.
>
> The time is now. Join us.

This call to action has much in common with historical examples of alternative journalism forms evident in both radical publications and what Christians *et al.* (2009) might also term 'facilitative' efforts in journalism. Britain's alternative publication from the 1990s, *Squall*, as one example, attempted to activate its readers as follows:

> Information is your weapon.
> The purpose of this magazine is to tool you up.
> With accurate information and positive inspiration.
> To expose hidden agendas and highlight new initiatives.
> Standing for cultural diversity, community and respect.
> To give fair voice to those who have none, have gone hoarse, or are frightened to speak.
> To battle for a better environment – countryside, urban and psychological.
> With no book, no badge, no flag and no anchoring affiliations other than the truth.
> . . . Arm yourself. (Malyon 1995)

The Guardian identifies that previous climate change coverage has left audiences 'feeling disheartened and disengaged' (March 15, 2015) and that the KIITG campaign is a real opportunity to engage, feel empowered and act by supporting the divestment aims and goals of the campaign. Audiences can, for example, donate to the campaign; contribute to letter-writing campaigns to the Wellcome Trust and the Bill and Melinda Gates Foundation to encourage them to divest of all fossil fuel investments; and take this technique to other major financial institutions, trusts, superannuation funds, etc. once divestment is accepted as an effective and achievable political goal.

The experiences of Guardian journalists

The Guardian presents, as part of its campaign, a series of 12 podcasts, recorded over the first five months of the campaign, which follow journalists on their journey to 'find a new way of telling the story' as directed by editor Rusbridger. These podcasts (see Data Reference list) record the conversations they have to find ways of overcoming journalism's 20-year struggle to report these issues in a manner that will empower, rather than disengage, its audience. Outgoing editor Alan Rusbridger, who saw KIITG as his final campaign, says in Episode 1 of the podcast series that climate change is 'clearly the most important story that we could be thinking about and yet you scan the daily newspapers and it's almost absent'. The podcasts reveal that conversations with Bill McKibben had influenced Rusbridger to begin KIITG to find a new way to report climate change. Rusbridger remembers a conversation with McKibben in Stockholm in 2014:

> [He told me] You guys have done fantastic reporting on the environment, on science, but this is no longer an environment story, this is a story where the science is settled. It's now all about politics and economics and if you don't mind me saying so, you're a bit old-fashioned in what you're doing.

KIITG is underpinned by Rusbridger's belief that journalism – a profession that has served him for his entire career – has failed on the issue of climate change. And he sees its failure directly related to the very nature of journalism as it is practised – climate change is an issue that 'does not change much from day to day', while:

> Journalism is brilliant at capturing momentum, or changes or things that are unusual. If it's basically the same story every day, every week, every year then I think journalists lose heart.

One of *The Guardian*'s journalists, who attended the first meeting at the newspaper to set up the KIITG campaign and to work out, journalistically, how they will carry it out, says news reporting has been 'really bad at changing the story where climate change is concerned':

> We carry on flogging a load of dead horses, and we are flogging them in exactly the same way with exactly the same whip, and it doesn't work. So we have to be constantly reinventing our story-telling capacity.

Over a five-month period, Rusbridger, his senior editors and journalists looked for new angles on the climate change story which are designed to engage the public more with the issue, and the podcasts record their deliberations. One of the key ideas they develop is to orbit their reporting around the central idea of divestment. The narrator to the podcast series, journalist Alex Krotoski, tells the audience in Episode 4:

> *The Guardian* wanted to focus [its campaign of divestment]. It picked two targets to ask to divest – not the worst kids on the block, nice liberal ones, a little like *The Guardian* – who are doing good stuff but shouldn't be in the messy world of oil, gas and coal.

In this, *The Guardian* staff are identifying the publication as a left-ish, progressive publication that is not radical but which, on this issue at least, is trying for social change. The conversations between journalists and senior editors evident throughout the podcasts suggest they are led by Rusbridger to develop an advocacy form of journalism which is campaigning for a particular outcome – for major financial institutions to divest themselves of fossil fuel investments. This campaign sits uncomfortably with some of the staff, and they are told it is not well-received by major news organizations in the United States who question *The Guardian*'s advocacy approach. Rusbridger says while news media sometimes take an advocacy stance on some issues – such as female mutilation – they often do so knowing there is little argument to be had *against* their stance. Climate change is different:

> Climate change for whatever reason is seen as a controversial subject and so therefore some people see that it's the duty of media to be objective and

impartial and not take a side and I think it has surprised some people that the newspaper has taken a side, and said 'we believe in the science.'

This type of approach leads *The Guardian* to develop a series of investigative stories around the major fossil fuel companies, to have their senior financial editor '"do the maths" to determine exactly what the cost of leaving fossil fuels in the ground would be', to canvas the range of views that different world religions have on climate change, and to run an interview with Shell CEO Ben Van Beurden who agreed to speak to *The Guardian*. The economics editor assigned to 'do the maths' says his major task was to write up what the world economy would look like without fossil fuels. He finds it would be an 'absolutely colossal economic transformation', with fossil fuel companies written down and their assets for future exploration and future profits written off. The economic impact of this would be as bad as or most likely worse than the sub-prime crisis which triggered the 2009 Global Financial Crisis. It would, according to *The Guardian*'s economics editor Larry Elliot, cause a massive stock market crash. Through the podcasts, we hear Elliot discuss the content of the article with Rusbridger and his fellow journalists and walk through the various scenarios that we might expect to play out in the case of fossil fuels staying in the ground. They rely for background information on their own analysis of the global economy and the monetary values of major fossil fuel companies, as well as the work of advocacy organizations such as carbontracker.org, which has previously carried out financial evaluations around climate change. The point of the discussions – and of the eventual article – is not to suggest that KIITG is an impossible campaign, but rather to understand the complexity of the issue and to expect that the changes being called for will occur over time. In this, *The Guardian* is trying to ensure they have covered the issue comprehensively; while some forms of advocacy journalism might ignore scenarios that challenge their overall approach, the KIITG editorial staff are ensuring they are fully aware of both the pros and cons of the campaign they are committing to.

At the end of the second phase of the KIITG campaign – the final weeks leading up to Rusbridger's retirement – the journalists re-evaluate the impact of the campaign. They recognize that not enough thought was put into their original notion to use divestment as the touchstone for the campaign, and they indeed received advice from the international Internet-based social change campaigner avaaz.org that divestment was not 'sexy enough' to gain mass public support. *Guardian* journalist James Randerson says the team is 'a bunch of journalists who do journalism, and don't really have in-depth knowledge when it comes to campaigning' (Episode 4, 'Risks'). Similarly, in the same podcast, the senior editor for strategy and partnerships thinks out loud:

As a news organization we have a really fast metabolism, so what we do today we can forget tomorrow, right? A campaign is certainly looking towards tomorrow but you're also bringing people along with you . . . there's a real feedback between ourselves and our readers that I think is new for us.

Overall *The Guardian* produced 200 stories about climate change during the first five months of the KIITG campaign, and those 200 stories attracted six million page views. They also had just over 200,000 people sign their petition to ask the Wellcome Trust and the Bill and Melinda Gates Foundation to divest of their fossil fuel interests, and 125,000 of those petitioners asked to be kept updated about the progress of the campaign. Both the Wellcome Trust and the Bill and Melinda Gates Foundation did not divest their fossil fuel interests – and gave various reasons for doing so, with the Wellcome Trust primarily arguing they had decided to stay 'in the tent' (so to speak) than to get out of the game altogether and divest. It is clear from the podcasts that Rusbridger, his senior editors and the journalists are disappointed they did not gain traction with the two main targets of the divestment campaign, although they do feel they have had a strong impact on public sphere debate around divestment of fossil fuel interests. Major international newspapers such as *The Financial Times* ran stories about divestment in which they interviewed major companies about their views on the issue. The campaign had also gained Rusbridger an audience with major financial institutions, politicians and international campaign organizations who had heard about the campaign.

The Guardian journalists observe that a major achievement of the campaign to date was to give readers a sense of agency. More radically, *Guardian* columnist George Monbiot said the newspaper would go after the fossil fuel industry through this campaign more than it had ever done before (Episode 2):

> Oil firms have to be made the pariahs of the world. They have to be stigmatized, they have to be reduced in their stature. . . . If we want to change the world . . . then we've got to actually deploy the measures that are going to change the world. And that's only going to happen through acting at the political level to lay down regulations which say, 'those fossil fuels are going to stay in the ground.' Everything else is prodding around on the edges of the problem, and not actually grasping that problem.

Concluding thoughts: Keep it in the Ground, facilitation and radicalism

The Guardian's KIITG campaign *did* find a new way to tell the story, although it may not have had the immediate impact that both Rusbridger and his team of journalists had hoped. They approached the issue of climate change in an advocacy manner, deciding that the science was now confirmed and that, as a result, they had a responsibility as a leading newspaper to stop covering climate change denialists and start looking for solutions. Within the team of journalists, there were differences about the tack to take; Rusbridger and many of the journalists advocated using known tools of journalism to simply dig further and more often and to provide original analysis, while commentators such as Monbiot were really looking for radical attacks on the fossil fuel industry in an effort to almost (or completely) demonize the industry for readers. *The Guardian* is not a radical newspaper, and the data examined here establish that well – indeed, it seems important to the newspaper that it is perceived as a progressive, liberal publication rather than an alternative or radical one.

This case study – examining newspaper content, journalistic processes and the stated motivations and proposed outcomes – suggests *The Guardian* continues to practise advocacy journalism on climate change. In this case, their advocacy *was designed to* directly encourage civic engagement and action by involving its audiences in the campaign for divestment of fossil fuel interests. There is no suggestion in either the content or in the bulk of the journalists' deliberations that the newspaper is advocating radical change to society, although it is calling for radical decisions from major financial institutions to remove one of their most profitable stocks – fossil fuels.

If we use our earlier concerns about the importance of journalism focusing on the *probable outcomes* of their work, this case study illustrates that *The Guardian* used the impact of their work as a key motivating factor – they wanted to see climate change taken seriously as one of the greatest threats to humanity; they wanted to exert a major public campaign to pressure financial institutions (i.e. capital) to withdraw investment from fossil fuels. More radical commentator George Monbiot pushed the argument to *Guardian* journalists that government legislation was also needed to give the campaign full effect – indeed he argued that the climate change debate so far had focused on *consumer behaviour* rather than *producer behaviour*, and this was creating a major gap in the efforts properly to address climate change. In short, there was a need for journalism to shift its focus from the seven billion consumers of fossil fuels to the few major corporations who keep 'digging up this stuff' and doing everything possible to maintain consumer demand (Episode 2, 'An angle').

In *The Guardian*'s KIITG campaign, we see strong evidence for the *facilitative* role of journalism, specifically a conscious mission to improve the quality of public life (as our Introduction outlined) and to promote active citizenship (Christians *et al.* 2009: 126). Civic democracy is the key concern of the facilitative role which sees the public as key *actors* who can support solutions and resolve public problems. The authors identify the facilitative role as something of a reformist rather than a revolutionary purpose for the media outlet, and this fits squarely with *The Guardian*'s outward persona. Let us consider the radical role for a moment though, as some data coming through *The Guardian*'s journalistic discussions revealed in the podcasts hint at a far more radical campaign – one which will shift social and political priorities in a major way. I refer specifically here to Monbiot's call to demonize the fossil fuels industry and to couple the focus on consumer behaviour with a commensurate focus on the behaviour of a few major fossil fuel companies. Our Introduction outlined that radical journalism aims to eliminate concentration of social power to achieve true equity and participation in decisions affecting society. It has sympathy with social movements representing marginalized or disenfranchised groups, and it takes the public engagement and activation that characterizes the facilitative role to a new level – it encourages changes 'in the core of existing social institutions' and supports change in the systems of communication (Christians *et al.* 2009: 179). Based on the case study, there is evidence that *The Guardian*'s KIITG fulfilled a facilitative role far more comfortably than a radical role, even though they relied on information from social movements such as 350.org and carbontracker.org that *do* appear to be fulfilling a more radical role. *The Guardian*'s campaign, then, was informed by more radical solutions but continued to operate within a reformist framework – looking for ways to improve the current condition rather than fundamentally to change it. The latter will be the task of more leftist publications, social media groupings and online media.

The KIITG campaign carries out important professional practice – one of our 'journalisms' – to find ways to make *addressing* climate change a focus of political and public action. Note here the focus is not on analyzing *The Guardian*'s 'coverage of' climate change but rather its attempts to 'address' climate change. It occurred within the structure of an independent publication, perhaps not a radical one, but certainly a contemporary, online, independent newspaper driven by its social responsibility. Indeed, this responsibility to deliver sound public information and tools to activate the public overrode any commercial imperatives that might have been an important marker of alternative journalism (Atton and Hamilton 2008; Forde 2011). The case study provided here suggests that forms of journalism focused on social change, even when centred on quite controversial issues within our political economy, can be successfully confronted outside the often-marginalized alternative and radical media mastheads.

This chapter's focus was on the place of digital media – indeed, all media underpinned by the Internet – in the modern journalisms which confront environmental issues. The case study of *The Guardian* illustrates that the KIITG campaign was made available to an Australian audience only because the economics to do so were delivered by online platforms. This is an important advance offered by digital forms. The impetus for the work, though, began with the traditional 'legacy' version of *The Guardian* in the U.K., and this demonstrates, as well as anything, that it is not the method of dissemination that drives change but rather the commitment and resources to do so, regardless of the platform.

How can such commitment and resources find a secure place in the media system? In concluding this book, attention turns to connections between public engagement, journalism and democratic media reform as a way forward.

References

Anderson, H. (2012) *Raising the Civil Dead: Prisoners and Community Radio*, Bern, Switzerland: Peter Lang.

Atton, C. (2002) *Alternative Media*, London: Sage.

——— (2003) 'What is alternative journalism?', *Journalism* 4(3): 267–72.

——— (2007) 'A brief history: The Web and interactive media', in K. Coyer, T. Dowmunt and A. Fountain (eds), *The Alternative Media Handbook*, Oxon and New York: Routledge, pp. 59–65.

Atton, C. and Hamilton, J. (2008) *Alternative Journalism*, London: Sage.

Bainbridge, J., Goc, N. and Tynan, L. (2015) *Media and Journalism: New Approaches to Theory and Practice*, 3rd edn, Melbourne: Oxford University Press.

Bennett, W.L. (2003) 'New media power: The Internet and global activism', in N. Couldry and J. Curran (eds), *Contesting Media Power: Alternative Media in a Networked World*, Lanham, MD: Rowman and Littlefield, pp. 17–37.

Bradshaw, P. and Rohumaa, L. (2011) *The Online Journalism Handbook: Skills to Survive and Thrive in the Digital Age*, New York: Longman.

Bruns, A. (2015) 'Working the story: News curation in social media as a second wave of citizen journalism', in C. Atton (ed), *The Routledge Companion to Alternative and Community Media*, London and New York: Routledge, pp. 379–88.

Cammaerts, B., Mattoni, A. and McCurdy, P. (eds) (2013) *Mediation and Protest Movements*, Bristol, UK: Intellect.

Carey, J. (1969) 'The communications revolution and the professional communicator', republished in E. Stryker Munson and C.A. Warren (eds) (1997) *James Carey: A Critical Reader*, Minneapolis: University of Minnesota Press, pp. 128–43.

——— (1989) *Communication as Culture: Essays on Media and Society*, New York: Routledge.

Christians, C., Glasser, T., McQuail, D., Nordenstreng, K. and White, R.A. (2009) *Normative Theories of the Media: Journalism in Democratic Societies*, Urbana and Chicago: University of Illinois Press.

Downing, J.D.H. (2003) 'The independent media centre movement and the anarchist socialist tradition', in N. Couldry and J. Curran (eds), *Contesting Media Power: Alternative Media in a Networked World*, Lanham, MD: Rowman and Littlefield, pp. 243–57.

Flew, T. (2008) *New Media: An Introduction*, 3rd edn, Melbourne: Oxford University Press.

Flyvberg, B. (2004) 'Five misunderstandings about case-study research', in C. Seale, G. Gobo, J.F. Gubrium and D. Silverman (eds), *Qualitative Research Practice*, London: Sage, pp. 420–434.

Forde, S. (2011) *Challenging the News: The Journalism of Alternative and Community Media*, Basingstoke, UK: Palgrave Macmillan.

Foxwell-Norton, K. (2015) 'Community and alternative media: Prospects for twenty-first-century environmental issues', in C. Atton (ed), *The Routledge Companion to Alternative and Community Media*, London and New York: Routledge, pp. 389–97.

Gerbaudo, P. (2012) *Tweets and the Streets*, London: Pluto Press.

Gynnild, A. (2014) 'Journalism innovation leads to innovation journalism: The impact of computational exploration on changing mindsets', *Journalism* 15(6): 713–30.

Habermas, J. (2006) 'Political communication in media society: Does democracy still enjoy an epistemic dimension? The impact of normative theory on empirical research', *Communication Theory* 16: 411–26.

Hutchins, B. and Lester, L. (2015) 'Theorizing the enactment of mediatized environmental conflict', *International Communication Gazette* 77(4): 337–58.

Hyde, G. (2002) 'Independent media centers: Subversion and the alternative press', *First Monday* 7(4). Accessed at http://firstmonday.org/htbin/cgiwrap/bin/ojs/index.php/fm/article/view/944/866.

Juris, J. (2005) 'The new digital media and activist networking within anti-corporate globalization movements', *The Annals of the American Academy of Political and Social Science* 597: 189–208.

Leaning, M. (2011) 'Understanding blogs: Just another medium?', in A. Charles and G. Stewart (eds), *The End of Journalism?* London: Peter Lang, pp. 87–102.

Lesage, F. and Hackett, R.A. (2013) 'Between objectivity and openness: The mediality of data for journalism', *Media and Communication* 1(1): 39–50.

Lister, M., Dovey, J., Giddings, S., Grant, I. and Kelly, K. (2009) *New Media: A Critical Introduction*, 2nd edn, London and New York: Routledge.

Malyon, L. (1995) 'Might not main', *New Statesman and Society* 8 (March 24): 24–6.

McKibben, B. (1989) *The End of Nature*, New York: Random House.

Meadows, M., Forde, S., Ewart, J. and Foxwell, K. (2007) *Community Media Matters: A Report on a National Audience Study of Community Broadcasting*, Melbourne, Australia: Community Broadcasting Foundation.

Platon, S. and Deuze, M. (2003) 'Indymedia journalism: A radical way of making, selecting and sharing news?', *Journalism* 4(3): 336–55.

Poell, T. and Borra, E. (2012) 'Twitter, YouTube and Flickr as platforms of alternative journalism: The social media account of the 2010 Toronto G20 protests', *Journalism* 13(6): 695–713.

Poulton, L; A. Purcell; E. Ochagavia; P. Wyse; D. Levene; B. Rinvolucri; F. Panetta and P. Boyd (2015) 'What is fossil fuel divestment and why does it matter? Video', *The Guardian*, March 23, 2015, from https://www.theguardian.com/environment/video/2015/mar/23/what-fossil-fuel-divestment-why-matter-climate-change-video?CMP=share_btn_tw

Radsch, C. (2011) 'Arab bloggers as citizen journalists (transnational)', in J. Downing (ed), *Encyclopedia of Social Movement Media*, Thousand Oaks, CA: Sage, pp. 61–4.

Rennie, E. (2006) *Community Media: A Global Introduction*, Oxford, UK: Rowman & Littlefield.

Rodriguez, C. (2001) *Fissures in the Mediascape: An International Study of Citizen's Media*, Cresskill, NJ: Hampton Press.

Sreberny, A. (2011) 'Social movement media in 2009 crisis (Iran)', in J. Downing (ed), *Encyclopedia of Social Movement Media*, Thousand Oaks, CA: Sage, pp. 497–9.

Stith, P. (2005) 'A guide to computer assisted reporting', Florida: Poynter Institute. Accessed at http://www.poynter.org/2005/a-guide-to-computer-assisted-reporting/69334/.

The Guardian (2015) '"Find a new way to tell the story": How The Guardian launched its climate change campaign', March 15, accessed from https://www.theguardian.com/environment/2015/mar/12/find-a-new-way-to-tell-the-story-how-the-guardian-launched-its-climate-change-campaign

Wolfsfeld, G., Segev, E. and Sheafer, T. (2013) 'Social media and the Arab Spring: Politics comes first', *The International Journal of Press/Politics* 18(2): 115–37.

Data references

Carrington, D. (2015) 'Guardian media group to divest its £800m fund from fossil fuels', *The Guardian (Australia)* (April 2). Accessed at http://www.theguardian.com/environment/2015/apr/01/guardian-media-group-to-divest-its-800m-fund-from-fossil-fuels.

The Guardian, Podcast series (2015) Episodes 1–12. Accessed at http://www.theguardian.com/environment/ng-interactive/2015/mar/16/the-biggest-story-in-the-world:

    Episode 1: Keep it in the Ground

    Episode 2: An angle

    Episode 3: The targets

    Episode 4: Risks

    Episode 5: Economics

    Episode 6: Psychology

Episode 7: Attacks
Episode 8: The U.S.
Episode 9: Religion
Episode 10: Shell
Episode 11: Investigations
Episode 12: Impact

Krotoski, A. (2015) 'Find a new way to tell the story: How *The Guardian* launched its climate change campaign', *The Guardian (Australian Edition)* (March 12). Accessed at http://www.theguardian.com/environment/2015/mar/12/find-a-new-way-to-tell-the-story-how-the-guardian-launched-its-climate-change-campaign.

Randerson, J. (2015) 'A story of hope: *The Guardian* launches phase II of its climate change campaign', *The Guardian (Australian Edition)* (October 15). Accessed at http://www.theguardian.com/environment/2015/oct/05/a-story-of-hope-the-guardian-launches-phase-two-of-its-climate-change-campaign.

Rusbridger, A. (2015) 'The argument for divesting from fossil fuels is becoming overwhelming', *The Guardian (Australian Edition)* (March 17). Accessed at http://www.theguardian.com/environment/2015/mar/16/argument-divesting-fossil-fuels-overwhelming-climate-change?CMP=share_btn_tw.

Conclusion

Media reform for climate action[1]

Robert A. Hackett

Climate crisis raises many questions about media and journalism that invite more attention from researchers and activists than can be addressed in a single volume. For example, how is the place of journalism in addressing global crisis affected by the digital mediascape's emerging characteristics, such as fragmented episodic news consumption? Can long-form journalism and approaches such as Peace Journalism assemble not only the necessary production time, skills and resources but also the attention and engagement of broad publics? What does a crisis and/or justice orientation mean for journalism education?

Beyond the media field, what about the shifting global political environment? In Europe, Asia and the U.S., we see a corrosion of commitments to equality or universal human rights, the rise of anti-immigrant parties exploiting the insecurity of those 'left behind' by neoliberal globalization, a mood of right-wing populist ethno-nationalism bordering on fascism, terrorism fuelled by religious fundamentalism, and a politics of pulling up drawbridges, building walls or (to mix metaphors) scrambling for the lifeboats instead of trying to prevent the ship from sinking. Climate change and other environmental threats, such as resource depletion, arguably accelerate this process, as does a global economic system extracting fossil fuels with increasing intensity and creating ever-larger 'sacrifice zones' and expendable people (Klein 2014). That depressing scenario is offset somewhat by the emergence of transnational civil society organizations, shifting public opinion, growing awareness and community place-based resistance to extractivism. But the side effects of climate change – mass migrations/displacement, growing competition for resources as basic as food and water – as generated by and mediated through dominant but contested institutions and templates such as militarism, nationalism, neoliberal capitalism and racism generate political conditions not conducive to the cultural resources needed to address the crisis: empathy, hope, solidarity, other-oriented ethics, political efficacy, civic trust and belief in the possibilities of collective action.

Thus, throughout this book, we have argued that climate crisis is not just a matter of environmental degradation but also of political and communicative capacity. In addressing that deficiency, journalism has an indispensable role. Clearly, journalism cannot do everything. We should not be taken as blaming news media, and still less working journalists, for inaction on global warming. The picture includes government policies, renewable energy technologies within the context of regulated carbon ceilings, and social movements that can pressure sluggish governments beholden to extractivist or fossil fuel–intensive industries and reluctant to challenge an economic system addicted to permanent growth and arguably engaged in 'creative self-destruction'(Wright and Nyberg 2016) – an insatiable dragon eating its own tail.

Journalism can, however, facilitate the kind of public engagement and mobilization that can create a bridge to political change. What would such journalism look like? The research literature and data presented in this book offer clues. Journalism can tell local stories, inspire community-level resistance and transformations, and amplify the counternarratives that can give meaning, direction and a sense of connection for people who are becoming active citizens. It could celebrate and normalize political action by ordinary citizens, tell success stories about climate politics and provide more concrete 'procedural knowledge' about how to take political action (Cross *et al*. 2015) – as well as supportive frames, such as the concept of climate justice. Conflict can be presented in ways that evoke outrage and mobilization rather than paralysis and cynicism by using the classic movement-building tactics of identifying grievances, enemies, allies and solutions. The key problem is *not* lack of information or climate denialism (although as Chapters Six and Seven indicate, there are strong denialist enclaves in Australian corporate media, probably more than in most other countries). Simply shovelling more data at people will not inspire them to act. Rather, the main blockage is a 'hope gap,' a discrepancy between the scale of the challenge and the sense of efficacy that ordinary people need as a basis for real engagement.

The aforementioned criteria generally match the facilitative and radical roles of democratic journalism. Our Australian case studies of the Keep it in the Ground campaign and independent media coverage of the Paris COP21 summit reinforce the growing scholarship suggesting that alternative media are particularly productive venues for journalism that takes on those roles. Alternative media's roles in community-building, collective identity formation and participatory citizenship are important and well-recognized. But it is arguably global crisis that renders urgent the need for alternative journalism and its most critical (in both senses of the word) functions – the formation and mobilization of counterpublics and of counternarratives. Counternarrativity entails filling in the gaps of dominant media accounts, finding the excluded voices and the dissonant facts that do not fit the official version, challenging repressive frames, providing new ways of making sense of contentious events and bringing attention to events and issues marginalized in the dominant media's topic agenda (Hackett 2016).

We thus conclude this book by asking: How can the scale of alternative journalism be ramped up to have greater resonance in global responses to climate crisis? What kind of economic and policy framework, consistent with notions of a free press and democracy, could offer an enabling environment for pro-climate media?

We offer a proposition that is foreshadowed but not developed in the existing research or political practice: *there is a coinciding agenda among climate action, alternative media and democratic media reform.*

Addressing market failure

In the Introduction, we outlined forces that make corporate media generally barren ground for productive climate communication; throughout, we argue that alternative journalism provides more fertile soil. Here, we go further, to suggest that the deficits of conventional climate news are not generally fixable within the framework of a privately owned, corporate-dominated, market-driven media system. Many of the factors that make for 'bad news' on the environment reflect longstanding critiques by civil society movements for the democratic reform of media structures and state communication policies that have emerged in the past few decades. The contradiction between media commercialism and such democratic values as equality and informed participation has motivated media reform organizations in a number of countries, such as Free Press and Media Alliance in the U.S., the Campaign for Press and Broadcasting Freedom and the Media Reform Coalition in the U.K., New Zealand's Coalition for Better Broadcasting, Australia's pro-public service media group ABC Friends (formerly Friends of the ABC), and OpenMedia in Canada. Here, I highlight the previously under-explored relevance of their critiques to climate politics journalism.

Consumer sovereignty, the idea that media give people what they want, is a key rationalization for a commercial media system. But while good work is done in some corners of the corporate media, they generally embed biases inimical to environmental communication. Their primary market is advertisers, not media consumers, generating pressure on content to be compatible with consumerism and to appeal to appropriate demographics – privileging affluent consumers over the less well-heeled, who are disproportionately the victims of ecological degradation. Market-driven media are not likely to give them a prominent voice. We have argued throughout that while climate crisis reinforces the urgency of journalism's ethic of truth-telling in the public interest, conventional journalism's 'regime of objectivity' incorporates practices that embed their own biases in favour of elite sources, prescheduled events and a debilitating he said/she said conflict narrative. And as Susan Forde argues in Chapter Seven, while the evolving online mediascape provides new opportunities for the distribution of alternative media, it is not likely to generate a green democratic technotopia under current conditions of development.

Media theorist Karol Jakubowicz (1999) argued that 'good broadcasting is a "merit" good – just as with education, training or health, consumers if left to themselves tend to take less care to obtain it than is in their own long-term interests' (p. 47). Much the same could be said of climate crisis journalism, given its complex and sometimes disquieting nature.

More recently, though, scholars are arguing that news produced by the media have some of the characteristics of a *public good* – a good that is difficult to commodify because it is non-rivalrous (one person's consumption of it does not detract from another's) and non-excludable (it is difficult to exclude 'free riders' who have not paid for it). Public goods are thus notably difficult to produce through market mechanisms (Pickard 2015: 213). Textbook examples are roads, streetlamps and national defence. Consumers can obtain a great deal of news (from environmental blogs to advertising-supported commuter dailies and urban weeklies to word-of-mouth) without direct charge. To be sure, crowd-funding through the Internet can help support individual bloggers and small-scale journalism organizations. But it is not clear that donations from already supportive individuals can expand climate journalism with the speed and scale needed. Besides, fund-raising already occupies too much of such independent media's energies, argue American media analysts Robert McChesney and John Nichols (2010).

Moreover, quality journalism provides *positive externalities* – benefits that accrue to people and society beyond the buyers and sellers directly involved in the transaction. Suggestively, comparative research by Toril Aalberg, James Curran and colleagues (2012) has demonstrated a positive relationship between the strength of a country's public service (as distinct from commercial) broadcasting system and the population's level of political knowledge and participation. And yet, although society benefits from such engagement, there are no obvious marketplace purchases whereby individuals can help pay for the costs of producing it.

Like other public goods generating positive externalities, journalism has never been fully financed by direct market transactions; in Canada and the U.S., journalism in commercial media has been in effect 'subsidized' by advertising for much of the past century. As we noted in the Introduction, that kind of revenue base is more likely to reinforce than to challenge environmentally destructive consumer culture. Moreover, that business model is in serious trouble as advertising and audiences migrate to the Internet. Given pinpoint target marketing made possible by online data collection, marketers no longer need to finance journalism as a 'free lunch' to attract audiences (Smythe 1981). Journalism is arguably becoming a case of 'market failure', a concept deriving from conventional neoclassical economics to describe a scenario:

> in which the market is unable to efficiently produce and allocate resources, especially public goods. This often occurs when private enterprise withholds investments in critical social services because it cannot extract the returns that would justify the necessary expenditures, or when consumers fail to pay for such services' full societal benefit.
>
> (Pickard 2015: 215)

The market failure of commercial media is compounded by their links with power, making the question of democratic communication unlikely to be addressed without a strong push from outside and below. In Canada, a landmark study of social and political inequality found multi-level links between the economic elite (senior executives and directors of major corporations) and media executives and owners (Clement 1975). Through old-boy networks, interlocking directorships, corporate-dependent media revenue streams and shared political perspectives, social backgrounds and career patterns, media and economic elites are fused into a cohesive corporate elite that wields financial and ideological power. Forty years on, in fossil fuel–exporting countries such as Australia and Canada, we need to update and expand this work to pay particular attention to the impact of fossil fuel corporations on Canada's economy, democracy and culture.

One relevant example of such influence is the reported backroom deal in 2014 between Postmedia, Canada's largest newspaper chain, and the Canadian Association of Petroleum Producers (CAPP). Labelled 'Thought Leadership', the proposed deal was to yield advertorials focusing on fossil fuel energy. Topics were to be directed by CAPP but written by Postmedia staff, with 12 single-page 'Joint Ventures' in 13 major Canadian newspapers (Uechi and Millar 2014). Tellingly, this deal was reported to readers not by Postmedia but by the small but ground-breaking independent *Vancouver Observer*.

Does such collusion between Big Media and Big Carbon really matter? Occasionally, I have heard social justice activists dismiss major media corporations as irrelevant 'legacy' media and conclude that media reform is unnecessary because activists have their own websites and digital networks. This view is problematic, given the continued reach and concentrated ownership and agenda-setting power of conventional media and, as argued in Chapter Three, the diminishing returns of staged media spectacles as a tactic for social movements. The corporate press may now share news dissemination with 'social media', but it continues to influence public policy discourse and agendas. Traditional media corporations have extended their presence onto the Internet, and they supply much of the information that is the basis for the blogosphere's opinion merchants. Conversely, the dependence of social justice groups on digital media makes communication policy principles (such as affordable, uncensored Internet access, as advocated by OpenMedia.ca) directly relevant to their work. And as noted in Chapter Seven, excessive dependence on social media leads social movements to reproduce some of the most problematic aspects of traditional media politics – spectacle and event orientation rather than elaborated explanations, solution-building and other aspects of pro-climate journalism, as discussed previously.

But where can such journalism be found? Mainly in news organizations independent of corporate ownership and commercial imperatives. We have cited examples from Australia – *The Guardian*, *New Matilda*, the *Green Left Weekly*, *Crikey*, and others. In Canada, amongst others nationally, rabble.ca has been offering moderately left-of-centre views since 2001. Vancouver is home to several outlets that emphasize pro-environment news. For example, *The Tyee* (the journalistic home of some of the respondents interviewed for Chapter Five) has offered investigative, analytical and solutions-oriented reportage on energy issues and much else since 2003 and competes well with the lacklustre corporate Vancouver dailies for readership. The *Vancouver Observer*, founded in 2006, provides bloggers and reporters with a platform on the environment and other issues, often telling stories from human interest and women's perspectives, such as a series on life in the oil sands epicentre of Fort McMurray. DeSmog.ca is an online news magazine 'dedicated to cutting through the spin clouding the debate on energy and environment', challenging climate science denialism and Big Carbon's public relations machinery.

In addition to shared grievances against corporate media, environmentally progressive alternative media are a second point of connection between climate activism and media reform, because such media struggle at the margins of the mediascape. They would benefit from public policy that offsets the systemic biases of market-driven media. And they can provide models of journalism that can influence larger media.

Policy support for more democratic media

The challenge is how to scale up the best practices and frames of such independent media to the point where they can influence 'mainstream' public discourse, given that the economic forces already noted not only favour commercialized media but also make sustainability difficult for independent, non-commercial, alternative media. Their content and demographics are unattractive to most advertisers. Politicians and other newsmakers often deny them quotes or access to news events. Alternative magazines do not enjoy much access to the semi-monopolistic distribution networks: you will not often see *Canadian Dimension* or Australia's *Red Flag* at supermarket checkout counters. They lack the cross-media resources to promote their websites competitively with corporate media. They typically rely on volunteer labour, grants and donations, and *de facto* subsidization from institutional sponsors such as foundations and trade unions. Even *The Tyee*, as one of Canada's most successful alternative independent news organizations, 'still struggles financially' (Skinner 2012: 42).

While alternative media are economically precarious, they add diversity to the media system, fill the growing gaps in local news and provide public voice for groups, topics and perspectives marginalized in dominant media. There is thus a strong democratic rationale for public policy support for alternative and independent media. What kind of policies? Many have been proposed and, in some cases, practised. How about charitable status for non-profit news organizations or a Citizenship News Voucher, as proposed by U.S. economist Dean Baker, whereby taxpayers can contribute $200 towards the non-profit news outlet of their choice (McChesney and Nichols 2010: 201–6)? Policymakers could facilitate the formation of trusts, like the one that publishes *The Guardian,* whose leadership for climate action was analyzed in Chapter Seven. They could achieve cross-subsidization within the media system to support non-profit public service media by charging small taxes on profit-oriented sectors, such as telecommunication services, cable television subscriptions, advertising or spectrum licences for commercial broadcasting. They could revitalize public service broadcasting, as the Friends of Canadian Broadcasting and Australia's ABC Friends advocate. Maintain Australia's leadership in community radio, and remove Canadian community access television from the grip of the cable monopolies, instead requiring them to fund multimedia community access centres as the Canadian Association of Community Television Users and Stations (CACTUS) recommends.

Such policies revolve around subsidies, incentives and infrastructure support. They need to be operated at arm's length from government to avoid political interference. Both Canada and Australia, after all, have ample experience in supporting public broadcasting, magazines, the arts, and television and film production through public institutions or programs. In the past decade, despite the bitter winds of a neoliberal political environment, media reform organizations in Canada and the U.S. have been able to mobilize people and win some regulatory victories, particularly in telecommunications. Starting almost from scratch, OpenMedia and Free Press each have attracted hundreds of thousands of supporters. A corporate-dominated media system is not inevitable or immutable.

These examples are illustrative, and no blueprint is intended here. Rather, the point is that media reformers and environmentalists could find common ground in favouring public policy support of alternative media and independent journalism as important pillars of both democracy and climate communication. More democratic media offer greater hope for a greener planet.

Note

1. Portions of the Conclusion derive from Hackett, R.A. (2016) 'Media reform and climate action,' (CCPA) *Monitor* (July/August): 40–5.

References

Aalberg, T. and Curran, J. (2012) *How Media Inform Democracy: A Comparative Approach*, New York: Routledge.

Clement, W. (1975) *The Canadian Corporate Elite*, Toronto: McClelland and Stewart.

Cross, K., Gunster, S., Piotrowski, M. and Daub, S. (2015) *News Media and Climate Politics: Civic Engagement and Political Efficacy in a Climate of Reluctant Cynicism*, Vancouver, BC: Canadian Centre for Policy Alternatives.

Hackett, R.A. (2016) 'Alternative media for global crisis', *Journal of Alternative and Community Media* 1(1): 14–6.

Jakubowicz, K. (1999) 'Public service broadcasting in the information society', *Media Development* 2: 45–9.

Klein, N. (2014) *This Changes Everything: Capitalism vs. the Climate*, Toronto: Alfred A. Knopf Canada.

McChesney, R. and Nichols, J. (2010) *The Death and Life of American Journalism*, Philadelphia: Nation Books.

Pickard, V. (2015) *America's Battle for Media Democracy: The Triumph of Corporate Libertarianism and the Future of Media Reform*, New York: Cambridge University Press.

Skinner, D. (2012) 'Sustaining independent and alternative media', in K. Kozolanka, P. Mazepa and D. Skinner (eds), *Alternative Media in Canada*, Vancouver: UBC Press, pp. 25–45.

Smythe, D. (1981) *Dependency Road: Communications, Capitalism, Consciousness, and Canada*, Norwood, NJ: Ablex.

Uechi, J. and Millar, M. (2014) 'Presentation suggests intimate relationship between postmedia and oil industry', *Vancouver Observer* (February 5). Accessed at http://www.vancouverobserver.com/news/postmedia-prezi-reveals-intimate-relationship-oil-industry-lays-de-souza.

Wright, C. and Nyberg, D. (2016) *Climate Change, Capitalism, and Corporations: Processes of Creative Self-Destruction*, Cambridge, UK: Cambridge University Press.

Index